A G R I C U L T U R E

Producer's Rationality and Technical Change

AGRICULTURE

Producer's Rationality and Technical Change

SIPRA DASGUPTA

ASIA PUBLISHING HOUSE
LONDON

Sipra Dasgupta (1930)

ISBN 0 210 22517 3

PRINTED IN INDIA

BY BRAHAM DUTTA BHARGAVA AT OXFORD PRINTCRAFT INDIA PVT. LTD., NEW DELHI AND PUBLISHED BY P.S. JAYASINGHE, ASIA PUBLISHING HOUSE, 447 STRAND, LONDON, W.C. 2.

To My Parents

PREFACE

IN INDIA, today, some of the basic problems of economic development centre round the agricultural sector. What are the possibilities of raising agricultural output? What is to be done to facilitate the growth of productivity in the agricultural sector? Are we fully utilising the cultivable land? Answers to these questions, in the last analysis, depend on our ideas about how the agricultural producers behave and *why they do what they are doing*. This aspect of the problem is of crucial importance in India where agricultural development is often based on the decisions taken individually by the multitude of producers.

The present study probes into the behaviour pattern of the agricultural producers in terms of economic rationality. It seeks to elaborate the nature of this rationality as expressed in the decision-making environment of the Indian agriculture. The risk elements, present in the farm-planning environment in India, have their impact on the existing technique in agriculture, on the selection of the crop alternatives, on the willingness of the farmers for new capital inputs—in short, the impact is on the economic rationality of the family farmers. In this context, the present study is a pioneering attempt at understanding one of the most difficult problems, namely, the producer's behaviour pattern. And it grew out of my doctoral thesis on "Producer's Rationality and Technical Change in Agriculture with special reference to India," submitted to the London University.

I would like to take this opportunity to express my deep gratitude to Mr. J. L. Joy who supervised my work at the London School of Economics. It is his constant help and guidance that enabled me to finish the work in its present form. For the discussion of specific problems, I am particularly indebted to Mr. P. Wegner, Mr. G. Morton, Mr. D. Throsby, Dr. D. Bose, and Dr. A. Bagchi. In connection with my study of the problems of technical change in agriculture in Japan, I often had to consult material in Japanese, and Professor R.P. Dore helped me with translations of the Japanese source material.

For my computational work, I am indebted to M.J. Garside of the London School of Economics and R.L. Cressy of the IBM Data Centre, London. Miss Diana Marshall-Say gave me concrete suggestions in editing the book.

I am grateful to the staff at the Cambridge University Library, India House Library, The British Library of Political and Economic Science, and The British Museum for their unstinted help.

The award of a Travel Grant by the British Council facilitated my visit to England. The Edwina Mountbatten Trust bore the cost of some translations from the Japanese. An award under the Endowed Research Time Scheme of the IBM, U.K., Ltd., enabled me to complete my computational work.

To my husband I am grateful for the frequent and fruitful discussions of my work which sustained my enthusiasm through all the hazards of a long research career.

SIPRA DASGUPTA

CONTENTS

LIST OF TABLES

INTRODUCTION

THE ROLE OF AGRICULTURE IN THE ECONOMIC DEVELOPMENT OF INDIA AND THE PROBLEMS TO BE INVESTIGATED

THE IMPORTANCE of agricultural development in the early stages of the economic growth of India arises from the following considerations: First, agriculture is the source of food. The increase in demand for food in India is related to a rapidly rising population. The rate of growth in population has been accelerating since 1920. The increase in population in the two decades, 1931-51, was about 82 million, while, during 1951-61, it was about 77 million. Taking account of the improving medical facilities, we can expect the present rate of growth—2.4 per cent per annum—to continue, unless a sharp fall in the birth-rate is brought about by a successful population control policy.

TABLE 1

POPULATION PROJECTIONS FOR INDIA: 1961-1976

(*in millions*)

1961	*1966*	*1971*	*1976*
438.0	419.5	554.7	625.1

SOURCE: Government of India, Planning Commission, *Third Five Year Plan*, 1961, Appendix C, Table 2, p. 751.

Apart from population growth, the increase in demand for food in India is due to a rise in the per capita income, given a high income elasticity of demand for food.

TABLE 2

PER CAPITA INCOME IN INDIA AT 1952-53 PRICES

(*Rs.*)[a]

1950-51	*1955-56*	*1960-61*
254	281	330

[a]The present exchange rate of rupee and sterling: Rs. 18 = £1.

SOURCE: Government of India: Planning Commission, *Selected Plan Statistics*, December, 1959, Table 9, p. 26.

Unless the agricultural output rises *pari passu* with the increase in population and in per capita income, in the early stages of economic growth, the shortages of foodgrains may create an inflationary barrier to further development. In India, a rise in foodgrain prices tends invariably to raise all other prices including the wage rate.

Theoretically, the growing demand for food can be met through imports. In practice, however, the ability of a newly developing country to import foodgrains may be restricted by two factors: (*a*) the scarcity of total foreign exchange resources relative to total need, and (*b*) competing claims on the available foreign exchange resources.

For instance, India's ability to earn foreign exchange resources is, primarily, restricted by the limited prospects of any rise in exports of the traditional product to the traditional markets, Secondly, it is restricted, by the inability to produce new commodities for export quickly and to create new markets for the traditional exports.

As regards the prospects of increasing exports of traditional products to traditional markets, the situation is described below.

Tea. It accounts for about one-fourth of the total merchandise export; there are possibilities of expanding tea output faster than the expected increase in internal consumption. Exports are limited, however, by the slow rate of growth in demand in the major markets which often impose high revenue duties on tea imports; the main competitors are Ceylon, and East Africa.

Jute Manufactures. The world market is stagnant due to the synthetic alternatives being produced in the major importing countries. The rising competitor is Pakistan. Industries in India are restricting output to below capacity level.

Cotton textiles. The two limiting factors on the increased exports of cotton textiles are the policies in many economically advanced countries to keep out cheap textile products and the growth of textile industries in other newly developing countries, e.g. Ceylon which previously imported textiles from India.

Agricultural products, e.g. foodstuffs, hides and skins and oilseeds. Such exports representing a small surplus over domestic requirements traditionally have declined due to the high rate of growth in internal demand. In fact, foodgrains are imported to supplement internal demand. Three to five per cent of total consumption was imported in the 1950's and their average value was Rs. 120 crores per annum.

New products, e.g. small engineering products, steel. Exports are rising mainly to other underdeveloped countries but are still not significant enough to make a difference to foreign exchange earnings. These will take time to grow sufficiently to make any important contribution to the balance of payment situation, although India has ample reserves of mineral resources, e.g. coal, iron ore, mica, manganese, aluminium, thorium, and labour is plentiful and cheap.

On the other hand, the importance of international trade in the case of India lies not merely in its size—traditionally moving round 7 per cent of the gross national product—but in the nature of the goods imported. These consist of materials already in use in some industry or service, e.g. metals, steel, oil, machinery for new development or replacements which cannot yet be substituted by domestic production, and marginal amounts of basic food-stuffs.

TABLE 3

INTERNATIONAL TRANSACTIONS OF INDIA, 1960-61 & 1961-66

Rupees in crores at 1960-61 prices

	1960-61	*Third Plan (i.e. 1961-66 Total)*
Debits		
1. Merchandise imports (c.i.f.)	1,122	6,350
2. Investment income paid abroad (net)	43	450
Total	1,165	6,800
Credits		
3. Merchandise exports (f.o.b.)	651	3,515
4. Government (net)	34	191
5. Other services (net)	31	191
6. Private donations (net)	37	198
7. Private capital inflow (net)	10	290
8. Official donations and borrowing (net)*	402	2,415
Total	1,165	6,800

*Including running down of external reserves.

SOURCE: W. B. Reddaway, *The Development of the Indian Economy*, Allen and Unwin, 1962, p. 180.

Secondly, the importance of agricultural development in India arises from its role as the source of raw materials for such industries as cotton textiles, sugar and jute. The failure of agriculture to fulfil increases in demand for such raw materials will either inhibit the growth of such industries or put new pressures on foreign exchange resources.

Thirdly, since nearly half of the national income originates in the agricultural sector (Table 4), any prospect of a sharp rise in national income is dependent on what is happening there.

TABLE 4

NATIONAL PRODUCT OF INDIA BY ORIGIN (at 1952-53 prices)
1950-51, 1955-56, 1960-61

Sector	*Amount (rupees in crores) and as Percentage of Total* 1950-51	1955-56	1960-61	*Increase (%)* *First Plan*	*Second Plan*
1. Agriculture and allied pursuits	4450 (49)	5230 (48)	6170 (46)	18	18
2. Mining	80 (1)	95 (1)	150 (1)	19	58
3. Factory establishments	590 (6)	840 (18)	1380 (10)	43	64
4. Small enterprises	740 (8)	840 (8)	1085 (8)	14	30
5. Construction	180 (2)	220 (2)	295 (2)	22	34
6. Commerce, transport and communications	1650 (18)	1875 (17)	2300 (17)	14	23
7. Professions and services	1420 (16)	1700 (16)	2100 (16)	20	23
Total	9110	10800	13480	18	25

SOURCE: Government of India: Planning Commission, *Selected Plan Statistics*, December, 1959, Table 9, p. 26.

A rise in output in the agricultural sector facilitates indirectly the growth of the non-agricultural sectors by creating a rapidly expanding market for industrial products—both consumer goods (e.g. cloth, sugar) and capital goods (e.g. chemical fertiliser, improved tools and machines). In India, this rise in the turnover of goods and services between the agricultural and the non-agricultural sectors will increase the flow of the marketed surplus of agricultural products to urban centres. Since the existing standard of living in rural areas leaves the basic food requirements of the rural population unsatisfied, the increased flow of the marketed agricultural products to urban centres may be feasible only if it is based on a rise in total output in the agricultural sector rather than on a redistribution of the "given" output in favour of the urban areas. Lastly, the importance of agricultural development in the initial stages of economic development in India arises from its rôle as a source of labour supply. In the traditional economies, agriculture absorbs most of the available labour force; the growth of new sectors necessitates a redistribution of labour.

In the early stages of economic growth, a rise in labour productivity in the agricultural sector facilitates the flow of labour from the farm to the non-farm sectors. Thus, the economic development of a country tends to be associated with a declining proportion

TABLE 5

SHARE OF DIFFERENT SECTORS IN THE TOTAL WORKING FORCE OF INDIA, 1950 51

	Numbers (in lakhs)[a]	*Percentage*
1. Agriculture and allied pursuits	1036	72.4
2. Total of mining factory establishments and hand trades	153	10.6
3 Commerce, transport, and Communications	111	7.7
4. Pofessions, services and administration	133	9.3

[a]Lakh=100,000

SOURCE: Government of India, *Final Report of the National Income Committee*, February 1954, p. 108.

of the total labour force remaining in agriculture. For instance, the number of workers in agriculture in Japan fell from 15,573 in 1878-82 to 14,613 in 1913-17, while the population grew at rates varying between 0.8 and 1.3 per cent annually. In the same period, acreage increased by 35% and land productivity,

$$\text{i.e. } \frac{\text{net output}}{\text{arable land}},$$

by 80%. There was, thus, a sharp rise in the productivity of labour, i.e. 2.6% annually, remaining in agriculture.[1]

In the case of India, there is no immediate prospect of a net withdrawal of labour from the agricultural sector, since the estimated increase in jobs in the non-agricultural sectors is not enough to absorb the whole of new additions to the labour force due to rise in population and since there is already a backlog of unemployment in both the rural and the urban areas.

In the absence of any State provision for the unemployed, individuals survive in India by doing some "work" or the other on the family farms in the rural sector or in the overcrowded services and trades in the urban areas—from shoe-polishing to carrying shopping baskets. Hence, the symptom of unemployment and poverty is visible more in the lack of "productive" employment opportunities for individuals than in open unemployment.[2] Even then there are some indicators of visible or open unemployment today. For instance, according to the estimates of the National Sample Survey, 4th, 5th and 6th rounds, the number of totally unemployed working force can be placed at 2.24 million persons;[3] another 21 million persons working 8 hours a week can virtually be treated as unem-

[1]In Japan, the exodus of the existing rural labour force towards urban centres was slow compared to other economically developed countries, e.g. the U.K., in their initial stages of growth. K. Ohkawa, H. Rosovsky, 'The Role of Agriculture in Modern Japanese Economic Development,' *Economic Development and Cultural Change*, IX, No. 1, Part II, October, 1960.

[2]See W.B. Reddaway, *The Development of the Indian Economy*, Allen and Unwin, London, 1962, pp. 24-5.

[3]The incidence of unemployment has been estimated to be 0.66% in the rural areas and 6.1% in the urban areas. In India, 30.4% of rural families belong to the category of "agricultural labourers" (i.e. 50% or more earners in families report hired labour as their main occupation); half of such families do not have any land and the rest own tiny plots of land.

ployed, while the number of people working less than 4 hours a day is 45 million.

The reports from the Employment Exchanges which are now being set up in urban areas also indicate the presence of unemployment among the existing working force.

TABLE 6

NUMBER OF PEOPLE REGISTERED WITH EMPLOYMENT EXCHANGES IN INDIA (1950-51 TO 1958-59)

(in thousands)

	1950-51	1951-52	1952-53	1953-54	1954-55	1955-56	1956-57	1957-58	1958-59	1959-60	1960-61
1. Number of applicants on live registers of Employment Exchanges (monthly average)	331	329	438	522	610	692	759	922	1133	1421	1606
2. Number of applicants placed in employment (monthly average)	28	35	30	16	14	14	16	16	19	23	26

SOURCE: Government of India, Planning Commission, *Basic Statistics Relating to Indian Economy, 1950-51 to 1961*, December, 1961.

In terms of the employment situation, India appears to be a special case in the world in that the agricultural sector, which already absorbs 70 % of the working force, has very little prospect of releasing some labour force from the land. On the contrary, by 1965-66, the agricultural sector is expected to absorb as many as 380,000 new workers in addition to its backlog of the already existing unemployed and the underemployed labour force.

An assessment of the future possibilities of the agricultural sector reveals the following characteristics.

Land supply. Most of the fertile land has already been brought under cultivation in India; the total land area is over $1\frac{1}{4}$

million square miles, of which over half a million (323 million acres)[4] is under cultivation. (See Table 7)

Low Technique. Though most of the fertile land in India is used, it is used imperfectly. The majority of the farm workers work less than five acres. In the fertile areas, the density of population is high and the holdings are even smaller (e.g. Bengal, Kerala). The agricultural technique is based on land and labour with very little use of capital embodying modern scientific innovations. The source of power is draft cattle. A comparison of the share of agriculture in national product (49%) and in employment (71.8%) in 1950-51 shows how inefficient the sector is.

Technological Possibilities. There are, however, unutilised possibilities of technological innovations in Indian agriculture, specially in those directions which can raise the productivity of land considerably.

TABLE 7

STATISTICS OF WORLD LAND USE BY REGIONS

(acres)

Regions	*Per cent of Total Area*				*per caput*	
	Under Cultivation	*Meadows and Pastures*	*Forests*	*Other*	*Cultivated*	*Other*
Far East	18	13	23	46	0.6	1.5
Near East	7	17	11	65	1.5	15.0
Africa	9	23	26	42	2.6	12.4
Latin America	5	18	48	29	1.3	7.4
North America	11	13	34	42	2.9	11.4
Europe	31	16	30	23	0.9	0.7
U.S.S.R.	10	17	39	34	2.6	9.0
Oceania	3	53	6	38	5.2	60.1
World	10	19	30	41	1.2	4.6
India	48	4	15	33*	1.0	0.6*

*Of which 20 per cent is reported as unused but potentially productive and 80 per cent as built-on area and uncultivable.

SOURCE: P.V. Sukhatme, "Food and Nutrition Situation in India," Part II, *Indian Journal of Agricultural Economics*, July-September, 1962, No. 3, p. 26.

[4]Commonwealth Economic Committee, *India*, H.M.S.O., London, 1963.

Irrigation. The water resources of India are estimated at 1,356 million acre-feet of which 450 million could be used for irrigation affecting some 175 million acres. On the eve of the First Plan, about 76 million acre-feet or 17% of the usable flow were being used, the area irrigated from all sources being 51.5 million acres. By the end of the Second Five Year Plan, the utilisation of water resources had gone up to 120 million acre-feet covering 70 million acres.

Fertiliser. In irrigated and high rainfall areas (potentially, nearly 2/3 of the total cultivable land) very substantial increases in yield could be secured by applying principal plant nutrients, specially nitrogen. Experimental farm data confirm the above possibility in different parts of the country. The use of chemical fertilizers per acre of arable land in India is now 1 lb. of nitrogenous fertiliser; the comparable figure for Japan is 225 lb. Domestic demand for ammonium sulphate reached 1 million tons by 1959, when indigenous production was only 395,000 tons. The annual expenditure on the foreign exchange account for the import of fertiliser plant and machinery is estimated at Rs. 400 million in recent years.[5]

Improved Seeds. The area covered by improved seeds in 1955-56 was estimated at 55 million acres. By 1961, four thousand Seed Multiplication Farms have been set up to provide seed for multiplication by registered growers. The demand from cultivators cannot be met in full.

To conclude, agriculture in India has to provide food for some 8 million more people every year; it has to offer more productive employment to some 62 million farm operators who have, at present, low productivity and low standards of living. Fresh virgin land is practically non-existent, the only alternative is to introduce technological innovations, e.g. irrigation facilities, fertilisers and improved seeds which can raise yields per acre of land sown with a large variety of crops (Table 8) over a total area of 323 million acres of cultivated land.

Social Implications. The future role of the agricultural sector in the national economy as discussed earlier implies, *from*

[5]Commonwealth Economic Committee, *India*, H.M.S.O., London, 1963, p. 52.

TABLE 8

AREA OF SOWN ACREAGE AND PRINCIPAL CROPS IN INDIA

1950-51, 1956-57, 1960-61

(Million acres)

	1950-51	*1956-57*	*1960-61*
Net area sown	293.6	322.5	—
Area sown more than once	32.4	45.9	—
Gross area sown	325.9	368.4	—
Total area under cereals	193.3	217.5	222.5
Total area under pulses	47.2	58.2	57.3
Total area under foodgrains	240.5	275.7	279.8
PRINCIPAL CROPS			
Foodgrains:			
Rice	76.1	79.8	83.0
Wheat	24.1	33.4	32.0
Barley	7.7	8.7	8.0
Jowar	38.5	40.4	42.7
Bajra	22.3	27.9	28.2
Maize	7.8	9.3	10.8
Gram	18.7	24.3	23.2
Oil Seeds:			
Ground-nuts	11.1	13.7	15.5
Sesamum	5.5	5.4	5.4
Rape and Mustard	5.1	6.3	7.1
Castor Seed		1.4	1.0
Linseed	3.5	4.2	4.4
Fibres:			
Cotton	14.6	19.9	18.9
Jute	1.4	1.9	1.5
Others:			
Sugar Cane	4.2	5.1	5.8
Tobacco	0.9	1.0	1.0
Tea	0.8	0.8	0.8

SOURCE: The Commonwealth Economic Committee, *India*, H.M.S.O., London, 1963, Table 40, p. 82.

the standpoint of society, that the existing cultivable land should be utilised intensively through the application of new capital inputs (e.g. irrigation, fertiliser, seed) and of labour to raise both output and employment, and that the existing cultivable land should be fully utilised to raise output and to generate employment as much as feasible.

It does not follow, however, that the above implications will necessarily be valid from the standpoint of individual decision-makers in Indian agriculture.

For instance, new technological innovations will be a feasible proposition only if the individual decision-maker finds them acceptable in terms of their implications for the uncertainty present in farm-planning (specially arising from the role of natural forces in agriculture). Again, the adoption of the new capital inputs by individual decision-makers will depend on their impact on farm income.[6]

Similarly, from the individual decision-maker's standpoint, the full utilisation of land may act as an economic objective only if land is the only "limiting" factor in farm-planning. Taking into account the average size of landholding (7.5 acres), it is traditional to regard land as the "limiting" factor on individual family farms in India. However, considering the pattern of the distribution of land ownership in India, it may be reasonable to assume that some individual decision-makers may not find land the only "limiting" factor in farm-planning. In such cases, the full utilisation of land may not be a feasible economic proposition. (See Table 9)

Problems to be investigated. The present study is concerned, in particular, with the following problems that arise in the context of the discussion so far: (*a*) How far optimum decisions of individual producers in Indian agriculture are affected by the presence of uncertain elements in the farm environment. (*b*) How far the introduction of new choice variables, e.g. chemical fertilisers, might be effective in raising farm output and income

[6]The increase in output is not necessarily the only way to raise farm income. For instance, the rise in prices of agricultural products due to shortages, government subsidies, a change in the cropping pattern (i.e. from low priced to high priced crops) can also raise farm income from the individual decision-maker's standpoint.

in the short period. (*c*) Given the existing land structure, how far the full utilisation of land is a feasible economic proposition for individual decision-makers.

Since the basic institutional pattern of agriculture is individually-owned family farms, the above problems can best be understood in terms of the economics of individual farm-planning.

TABLE 9

DISTRIBUTION OF LANDHOLDINGS IN INDIA BY OWNERSHIP AND OPERATION—CROP YEAR 1953-54

Size Group	*Ownership Holdings*		*Operational Holdings*	
	No. of Households (% of total)	*Area Owned (% of total)*	*No. of Households (% of total)*	*Area Operated (% of total)*
(acres)				
0.00[1]	23.09	0.00	10.96	0.00
0.01-2.49	38.15	6.23	45.19	5.78
2.50-4.99	13.49	10.09	15.08	10.02
5.00-9.99	12.50	18.40	14.19	18.56
10.00-24.99	9.17	29.11	10.36	29.22
25.00-49.00	2.66	18.63	3.12	19.54
50.00 or above	0.94	17.54	1.10	17.08

[1]Landless.

NOTE: Nearly 75 per cent of the rural households (74.73 %) have land less than the national average of landholding, i.e. 7.50 acres and own 16.32% of the total land available; 12.77 % of households own 65.28% of the land in holding above 10 acres.

SOURCE: *India National Sample Survey*, 8th Round, July 1954-April 1955, Reports 10 and 30.

CHAPTER I

THE CHOICE OF METHOD IN THE PRESENT STUDY

A NATURAL scientist, undertaking a scientific investigation, can usually start from a generally accepted identification of the problem situations as well as a theoretical structure or framework in terms of which the problem is to be analysed. In the social sciences neither can be taken for granted. In the field of agricultural economics in underdeveloped countries, these difficulties are more than usually acute.

In this study, we present a generalised conjectural statement regarding the crucial factors influencing economic decisions taken by individual decision-makers in Indian agriculture. Any attempt at formulating generalised statements in the social sciences implicitly assumes "regularity" in behaviour patterns. Since without such "regularity" being present in sufficient degree, any production planning would have been impossible, we take it as a simple "fact" of life. On the other hand, formulating a generalised statement does not mean attempting to formulate a universal truth but only a statement with actuarial significance. Secondly, we state explicitly the assumptions (i.e. "basic statements" or "highest level hypotheses") [1] on which the generalised statement is based. Thirdly, we deduce from the generalised statement implications as to how decision-makers behave in particular situations. Fourthly, we try to refute the individual statements in terms of the actual behaviour pattern of the producers concerned. "Predictions" from a hypothesis can both be valid for what has happened in the past (i.e. to give a logical structure to what has already happened) and for what is expected to happen in future. We shall try to refute our "hypothesis" in terms of the already accumulated empirical data. Lastly, we shall try to indicate directions in which new evidence capable of refuting our hypothesis might be collected.

[1] R.B.Braithwaite, *Scientific Explanation : A Study of the Function of Theory, Probability and Law in Science*, Cambridge University Press, 1953.

Since the basic principle underlying the scheme of study outlined above is that the hypothesis under investigation should be capable of being refuted[2] and since the refutability of any hypothesis implies the realism of assumptions on which it is based, our present study will involve some discussion of the "realism" of the assumptions involved. Such a discussion, however, certainly does not constitute an attempt at "verifying" the hypothesis by a back door.

It is sometimes suggested that the assumptions of a hypothesis need not be realistic,[3] since the verification of a hypothesis can be done in terms of the realism of the conclusions drawn from it. But the verification of "conclusions" drawn from a hypothesis in n cases does not constitute the verification of the hypothesis for $n+1$ situations. On the other hand, the realism of the assumptions is a necessary condition for being able to subject a hypothesis to tests of refutability. Suppose, the hypthesis is, "*A* implies *B*." "*A* implies *B*," can be falsified by finding a situation of "*A* and *not B*." If *A* is an empty set, *not A* is always true. Then the hypothesis "*A* implies *B*" is not refutable.[4]

The choice of particular assumptions in our present study presents two problems. (*a*) The choice of particular assumptions always involves an arbitrary element, since they are, in general, derived not only from observations but also from other initial hypotheses which for the study in question are taken as given.

[2]Cf. Popper: "A theory which is not refutable by any conceivable event is non-scientific. Irrefutability is not a virtue of a theory (as people often think) but a vice." Karl R. Popper, *Conjectures and Refutations, The Growth of Scientific Knowledge*, Routledge and Kegan Paul, London 1963, p. 36.

[3]See Milton Friedman, *Essays in Positive Economics*, the University of Chicago Press, 1953.

[4]*Truth Table*

A	*B*	*ACB*
T	*T*	*T*
T	*F*	*F*
F	*T*	*T*
F	*F*	*T*

T = True

F = False

ACB = "*A* implies *B*"

(*b*) The economic behaviour of an agricultural producer may be influenced by many non-economic variables, (*i*) physical factors (e.g. climate), (*ii*) biological factors (e.g organic drives), (*iii*) psychological factors (e.g. attitudes, desires), (*iv*) legal factors (e.g. forms of property relations), (*v*) sociological factors (e.g. social cohesion or social role).

In the present study, the choice of assumptions will be guided by the following considerations.

First, the need to minimise the number of assumptions arises from the fact that the larger the number of assumptions made in a study, the greater is the possibility that all conceivable conclusions can be "justified" and hence, the hypothesis becomes incapable of being refuted. Secondly, to make such assumptions only which have a direct bearing on the implications drawn, the "relevance" is determined by the nature and the scope of the present study which does not pretend to be an exhaustive analysis of Indian agriculture. The hypothesis of the present study can thus be stated: the individual decision-makers in Indian agriculture allocate their resources rationally among alternative uses in farm-planning, guided by their own economic interests.

More specifically, an individual decision-maker in Indian agriculture seeks to raise his family income as modified by two factors, namely, the protection against natural hazards and the safeguard against the loss of the soil fertility over a period of time, within his given environmental constraints. Constraints imposed by social environment, i.e. factors which are given, so far as an "individual" decision-maker is concerned, are: (*i*) the choice variables by operating on which the individual can maximise his family income are limited by the state of the economic growth and by the state of the scientific knowledge in the country concerned, (*ii*) legal relationships between producers and the land which may inhibit the freedom of action of the individual decision-maker seeking to maximise family income, e.g. restrictions on the choice of crops, or on the scale of operations often implicit in the feudal types of land relation structures. Decision Theory has emphasised the need for a global approach to rational behaviour, i.e. what is important is not whether an individual behaves consistently in a given environment but whether he sets his environment rationally. Although "changing the environment" is an important aspect of man's behaviour pattern, it cannot be fruitfully discussed in the present context. The problem how an individual makes decisions *in a given environment* and how he takes actions *to change a given environment* are conceptually two different problems. For example, a peasant-producer in India who is otherwise "normal" and "rational" may fail to set his environmental limits rationally owing to a combination of factors. First, the peasant-producer may not simply know that there is any better alternative environment than that within which he himself operates. Secondly, the peasant-producer, instead of questioning his given environment, may simply expect the unfavourable effects of the present situation to be temporary. It is difficult to say, *a priori*, when and how an individual makes the decision as to whether an unfavourable state of affairs is a temporary phenomenon, or, a permanent feature of a given environment. Thus, the study of man's "rationality" in changing his environment can be built up only in terms of a general equilibrium model, taking into account all the variables affecting man as a living organism, e.g. economic, political, sociological, ethical.

We have drawn two implications from the hypothesis as formulated above. First, the existing husbandry practices and the choice of the cropping pattern (i.e. operation with the given choice variables) are determined by the above economic motivations subject to the constraints. Secondly, any new change or innovation in the cultural practices or in the cropping pattern (i.e. representing a feasible change in choice variables) is assessed by individual decision-makers in Indian agriculture in terms of its implications for the economic objective and the constraints on their achieving it.

We consider below certain objections that may be raised against our formulation of the economic objective of the agricultural producers in India.

It is often stated that the agricultural producer in India is not oriented towards any economic motives and that his conduct is rooted in Indian thought and philosophy which are essentially idealistic and other-worldly in nature.

We could not accept this argument for the following reasons. In the first place, the commonly accepted view that Indian philosophy and religion are essentially other-worldly does not appear to be borne out by scholarly study.[5] One important strand in Hinduism, for instance, is strongly oriented towards "action"—action in this life. The concept of Karma extols "action" while renouncing the involvement in the fruits of "action." Secondly, the sentiments and behaviour of an Indian peasant in an Indian village are far less likely to be influenced by philosophic thought than by religious rituals and social sanctions affecting him directly. These are frequently oriented towards material interests. In any part in India, for example, a Hindu peasant family offers daily prayer to the image of God as the source of Wealth (Lakshmi) and the source of Business Acumen and Success (Ganesha). In village life, material wealth enjoys such great social prestige that instances of wealthy men going up the caste ladder are not uncommon.

Regarding the social institutions it may be pointed out that the caste system does not enjoin a man to neglect his profession; on the contrary, it is the "duty" of one belonging to an agricultural caste to make a success of his occupation. Similarly, caste sanctions against field work do not touch the agricultural castes who are expected to work hard but only men belonging to the highest caste, i.e. the Brahmins, who hardly form two per cent of the total population in a village, and are generally religious men (i.e. living on religious endowments), and big landowners or merchants.

An argument closely related to the one considered above suggests that agriculture in India, is a "way of life" rather than an occupation for agricultural producers.

Available evidence, on the contrary, tends to show that the agricultural producers in India regard agriculture as a business pro-

[5]Das Gupta, for instance, reminds us, "The old civilisation of India was a concrete unity of many-sided developments in art, architecture, literature, religion, morals and sciences so far as it was understood in those days. Much harm has already been done by the circulation of opinions that the culture and philosophy of India was dreamy and abstract." S.N. Das Gupta, *A History of Indian Philosophy*, I, Preface Cambridge, 1932.

See also, A. L. Basham, "Modern Historians of Ancient India," *Historians of India, Pakistan and Ceylon*, ed. C.R. Philips, Oxford University Press, 1961,

position, a means of earning a livelihood. Whenever, there are better alternatives present, the agriculturists have shown a strong preference for giving up agriculture as an occupation. The mobility of the rural labour willing to work at the current wage rate in the non-farm sectors, the migration of rural labour to cities in England to take up industrial jobs[6] all tend to support this view.[7]

A better assumption is that an agricultural producer in India accepts agriculture as a means through which income can be earned. The high price elasticity of supply[8] when competing crops are present, the long-term shift of land and labour from food crops to cash crops enjoying favourable price ratios (Refer to Table 10) and, in particular, the rapid spread of new cash crops (jute, cotton, sugar-cane, ground-nuts, etc.)—all corroborate the present assumption. The adoption of improved production processes such as the intricate pattern of rotational practices, delicate adjustments of the timing of sowing and harvesting to rainfall and temperature variations, the highly developed legume cultivation, the use of a wide range of manurial inputs (e.g. cow-dung, oilcakes, river silt, etc.) which resulted in increased output within the given constraints of the state of knowledge and technique in society, all strengthen the validity of the above hypothesis.

Sometimes, it is argued that, even if an agricultural producer may be capable of aspiring after higher family incomes in principle,

[6]Recent sociological investigations into the occupational pattern of the Indian immigrants in England show that the majority of them belong to agricultural castes and have come straight from villages in Gujarat, Punjab or Assam. A small minority of the immigrants belong to craftsmen who depend for livelihood on agriculture and to Brahmin castes with rural origins. See Rashmi Desai, *Indian Immigrants in Britain*, Oxford University Press, 1963, pp. 14-15.

[7]Snow reminds us, " ... in fact, in any country where they have had the chance, the poor have walked off the land into factories as fast as factories could take them." See C.P. Snow, *The Two Cultures and the Scientific Revolution*, The Rede Lecture, 1959, Cambridge University Press, 1962, pp. 24-5.

[8]Two works on this problem deserve special reference: (*i*) The Statistical Institute, *Sankhya*, Calcutta, 1 May 1934 ; (*ii*) Dharam Narain, *The Impact of Price Movements on Areas Under Selected Crops in India, 1900-1939*, Cambridge University Press,

TABLE 10

DATA RELATING TO FOOD AND CASH CROPPING IN INDIA 1893/94

Years	All-India estimates of (Gross) Food crop production ; and ratio between non-food crop to food crop output			
	Average Annual Production of food Crops[1] *(million tons)*	*Non-food*[2] *to food crop output ratio*	*Indexes of average annual crop output* *food*[3]	
1893/94-1895/6	73·9	.22	100	100
1896/97-1905/06	71·5	·24	96	105
1906/07-1915/16	74.0	.28	99	126
1915/16-1925/26	73·3	·32	98	142
1926/27-1935/36	69.6	·41	94	171
1936/37-1945/46	69·3	·44	93	185

[1]Include rice, wheat, barley, jowar, bajra, ragi, maize, gram, "other food grains," and pulses.

[2]Include linseed, sesamum, rape-seed, mustard, groundnut, castor, sugar-cane, cotton, jute, indigo, coffee, tea, and tobacco.

[3]The indexes are based upon aggregates weighted by the average price of cash crops for the period 1924-25 to 1928-29.

SOURCE: George Blyn's estimates used by Dr. Thorner in "Long Term Trends in output in India," *Economic Growth: Brazil, India, Japan*, ed. S.S. Kuznets, Duke University Press, 1955, pp. 121-2.

he fails to do so in practice because of his low marginal rate of leisure-income substitution at the current level of income. A high relative preference for leisure has been sought to be inferred from various types of considerations.

It may be argued, for instance, that since in the traditional life in India wants are narrowly defined, a peasant-producer, once such wants are satisfied, can reasonably be expected to have a high marginal rate of leisure-income substitution. But known traditional wants of food, clothing, shelter, etc. remain unsatisfied today in the case of the majority of the agricultural families in India. This is evidenced by the econometric investigations into income-consumption patterns of rural families, in their high income elasticity of demand for innumerable indigenous dairy products, major cereals, coarse cloth, fat and sweet products.[9] Moreover, even if the known wants in a country prior to industrial growth are narrowly defined, compared to a modern industrial society,[10] such wants are much more varied than mere physiological needs and are hierarchical. Peasant producers in India are also influenced by the consumption patterns of their neighbours and want to acquire status symbols (e.g. ornaments, religious ceremonies) as acknowledged by people familiar with village life. Even in the backward Indian economy, some industrial growth took place over the last few decades creating new wants, e.g. mill-made cloth, mill-made sugar, tea, entertainment such as cinema.

The majority of the rural families are not only poor but they are also conscious of their poverty. That there is such dissatisfaction with the low standard of living — now often vocal—has been personally witnessed by the author herself. The success of conducting the electioneering in Indian villages in terms of the concrete promises of improvements in village life also testifies to the awareness of the rural families as to their poverty.

Secondly, it has been suggested that because of the climatic and similar factors, rural families have little capacity to work and to make efforts to raise family income.

It appears to be more reasonable to argue, however, that the working time of family members on individual farm holdings is, in practice, determined by the demand factor, i.e. labour time required on family holdings as determined by the condition of soil and rainfall, cropping pattern, the size of holding, etc. For example, agricultural research workers have often pointed out the scrupulous care of the land taken by a peasant family, the timely performance of tasks such as sowing, weeding and harvesting. Again, recent farm investigation in all parts of the country indicates that on the

[9]Ganguli and others, *Studies on Consumer Behaviour*, Indian Statistical Institute, Asia Publishing House, 1960.

[10]Cf. Mellor, J.W., "The Use and Productivity of Farm Family Labour in Early Stages of Agricultural Development," *Journal of Farm Economics*, August 1963.

small farms with plentiful family labour, the marginal unit of labour time applied to cultivation, given the state of technique, is associated with zero or near zero return. In such families, the so-called leisure-preference merely reflects the involuntary "idleness" of family members, working within the given technological constraints.

On the other hand, the large producers who have been found traditionally not to push the application of labour time to the point where its marginal product is zero, are motivated more by simple calculations of equating marginal revenue to marginal cost (since "hired" labour forms a significant proportion of the farm labour force) than by some "inherent" leisure-preference function. At the same time, the large producers also show initiative in branching out into non-agricultural occupations such as money-lending, trading , etc. to create new sources of family income, which may be taken to indicate for large farmers a higher expected return from extra "efforts" in such occupations than that in agriculture.

Sometimes, the observed phenomenon that the agricultural workers on family farms take rest in afternoons rather than working in the field is posed as the evidence of a high preference for leisure at the current income level. Yet, a working time of 10 to 5 is not of any special virtue; a typical peasant family shows great flexibility in the choice of working time in order to suit the climatic conditions. The author has frequently seen peasants avoiding the midday sun and working late at night or a whole family working continuously in order to complete the *Kharif* sowing which is a race against the monsoons. The point at issue is borne out by the experience of Sir Albert Howard, Director of the Experimental Farm at Indore.

The Indore Institute started with a 10-hour day which was the general practice in India; however, during the hot months of April, May and June, both men (members of peasant families) and cattle were found to require more protection from the hot sun. An experiment was made by reducing the working time during three hot months to six hours daily, four hours in the morning and two in the afternoon, and six hours rest from 10 a.m. to 4 p.m. The result was such a marked improvement in the health and well-being of men and animals and in the speeding up of work, that the working hours, all the year round, were reduced from 10 to 7-1/2 hours and 6 hours in the hot months. (This refers to time actually at work in the fields). This rule prevailed except for about a week at the sowing time of the monsoon crops when there is an intense race against Nature as realised by all the parties concerned.

Howard writes that, "for three years the agricultural operations of the Institute have been conducted on the short hours system. The result has been successful beyond all expectations. The miracle of speeding up Indian labour has been achieved and shorter hours have led not only to contentment but also to increased output of work."[11] Thirdly, a more sophisticated approach to the problem is sought to be based on the concept of the "level of aspirations" borrowed from the psychologists.[12] According to this view, Indian peasant-producers are poor because their aspiration levels are low. The aspiration level thus defined as the ex-post satisfaction level giving a solution for each act implies "what you get you aspire after." In this popular sociological version, the concept of "aspiration level" merely refers to a static equilibrium point and does not indicate either why an individual makes any choice at all or how he can make the optimum choice. A more satisfactory interpretation of the concept of the "aspiration level" might be to compare it to the Ricardian concept of Subsistence Wage rate which includes both the physiological and the conventional element. Since conventional needs adopt to the past consumption level, the subsistence wage rate is a moving point over a period of time. In this sense, the "aspiration level" might be regarded as the lower boundary of the utility function, to use the economic jargon. It is

[11]Albert Howard, "An Experiment in the Management of Indian Labour," *The International Labour Review*, Geneva, Vol. XXIII, No. 5, May, 1931. It might be noted that this success may have been facilitated by certain other measures which Howard introduced, namely, clean water supply, housing facilities, some free medical benefits, regular and honest payment of wage in cash, full freedom of the workers to spend their income as they like and where they like (i.e. no "farm shops" at the Institute where workers have to make their purchases.)

[12]See S. Siegel, "Level of Aspiration and Decision-Making," *Psychological Review*, Vol. 64, No. 4, 1957.

conceivable that in a particular place, at a particular time, the behaviour pattern of an individual may turn on this particular element in his utility function; for example, a peasant-producer concentrating on satisfying the family's food requirements after a famine, or, a landless labourer concentrating on meeting primary needs of the family after he gets land for the first time, or, a wage earner satisfying the family's needs after a long period of unemployment.

One important aspect of the concept of "aspiration level" emphasised by the psychologists is often neglected in its popular versions. The psychologists emphasise that the "aspiration" level tends to rise from point to point when an individual finds it easy to satisfy his level of aspiration at a point and that it tends to fall when the individual cannot do so. For instance, "apathy" and "hopelessness" sometimes reported to have been witnessed among the poorer strata of peasant-producers (specially among the landless agricultural labour families in India) may simply be the expressions of their "psychological withdrawal," i.e. zero aspiration level due to their continued failure to achieve their aspiration levels in the given socio-political environment rather than to some inherent peculiarities. It must be emphasised here that we cannot generalise from these particular investigations among particular categories of agricultural families as to the psychological states of the other categories of agricultural families (e.g. land-owning peasant families) or even of the former over a long-time horizon.[13]

As regards the possibility that the raising of family income gets modified in Indian agriculture by the role of natural factors, it may be regarded as devoid of any special significance, since natural forces play an important role in any agricultural system.

Our contention that natural forces play a specially significant role in Indian agriculture is based on a number of considerations. Nature affects agricultural operations in India through a large

[13]In recent psychological investigations, varied utility functions have been emphasised. See M. Dresher (ed.), "Contributions to the Theory of Games, Annals of Mathematical Studies" III, No. 39. 1957.

number of factors.[14] Nevertheless, we shall concentrate mainly on rainfall, i.e. a minimum requirement of agricultural operations. The rainfall in India is dependent on the behaviour pattern of the monsoons. While most of the subcontinent receives the maximum rainfall during the south-west monsoon, some parts of the country receive rainfall from the north-east monsoon in the winter months as well.

Over the country as a whole, the rainfall distribution is extremely uneven, e.g. a narrow strip on the western coast receives more than 80″ of rainfall, while the whole of the Deccan plateau lying on the east of the Western Ghat receives only 20″-40″ of rainfall. The agricultural crop pattern changes with the abruptly varying rainfall pattern over the land surface. For instance, the average rainfall at Lonavla on the edge of the Western Ghat (2,100 ft.) is 166.74; 20 miles further east at Vadgaon it is 41.19″, and at Poona (1,850 ft.) another 20 miles east, it is only 18.23″.

The rainfall in India is not only unevenly distributed over the country but it is extremely uncertain within one particular agricultural season, as well as over different seasons. Dr. Harold Mann concluded, on the basis of his study of the time series data of rainfall,[15] that it is not possible to make forecasts about the total rainfall for one season by observing the behaviour of the monsoons in the beginning of the season. An individual producer, in many parts in India, starts his sowing at the beginning of the rainy season without any knowledge how the rainfall will behave over the whole

[14]Dr. L.A. Ramdas, Agricultural Meteorologist, enumerated the major weather abnormalities with adverse effects on crops, as follows:

(a) Excessive rains
(b) Scanty rains
(c) Untimely rainfall
} Connected with the nature of the monsoons

(d) Storms, cyclones, depression (coastal regions)
(e) Thunder storms, dust storms and hail storms (all over the country
(f) Cold waves with frost (North India)
(g) Heat waves (North and Central India)
(h) Excessive and defective insulation
(i) High winds

[15]Harold Mann, *Rainfall and Famine—A Study of Rainfall in the Bombay Deccan, 1865-1938*, Indian Society of Agricultural Economics, Bombay 1955.

sowing time, specially in regions generally having less than 40″ of rainfall. He starts his sowing often without knowing definitely which particular crops are going to be sown, since he often adjusts his choice of crops to the distribution and the amount of rainfall in a particular season; in between sowing and the maturing of the crop, he remains in ignorance about the final outcome; when the harvesting starts he works with his simple tools without being sure whether an untimely shower will destroy his ripened crop or not.

In addition to the erratic behaviour pattern of the monsoons in general, certain abnormalities are inherent in its very nature. Such abnormalities are caused by seasonal perturbances, an abundance of which results in "floods" in a certain area, a scarcity results in "drought" in another, while the lack or the excess of such seasonal perturbances cause a large variation in the occurrence of rainfall regionally as well as seasonally as discussed earlier (Refer to Table 11)

Generally, when the rainfall in a year is roughly more than twice the mean deviation in a series of years, that year is designated as a year of "flood" or "drought", according as the departure is positive or negative. Although there is no regularity in the distribution of "floods" or "drought" over time or space, it is customarily assumed that the total number of abnormalities are the maximum where the rainfall is near about or below 40″ and that the frequency of abnormalities goes down when the rainfall is above 40″.

The economic loss due to the rainfall variations involves not only the loss of crop in real terms but also the death and the physical impairment of the labour force and often, the physical destruction of "cattle" (i.e. the most important element of "fixed" capital besides land). For instance, Dr. Harold Mann, Director of Agriculture, Bombay, calculated at constant price, the value of agricultural production in one district of Ahmednagar to be Rs. 1,37,968,000 in 1917-18 and Rs. 38,858,000 in 1918-19 (a year of drought).

In conclusion, we find that while the mechanism which causes the onset of monsoon, its continuation and final withdrawal has not yet been fully understood, the backward agricultural technique, based

TABLE 11

MEAN RAINFALL AND ITS VARIABILITY IN INDIA

Divisions	Rainfall[1] (June-September) (inches)	Rainfall[2] as % of annual rainfall	Coefficient of Variability (%)
(1)	(2)	(3)	(4)
1. Assam	68.1	66	10
2. Bengal	55.1	74	6
3. Bihar and Orissa	42.6	82	11
4. U.P.	33.9	88	21
5. Punjab	14.7	76	31
6. Rajasthan	19.5	91	31
7. Bombay (inc. Gujarat)	40.7	89	19
8. Madhya Pradesh	43.5	88	15
9. Madhya Bharat	35.6	91	20
10. Hyderabad	28.4	80	22
11. Mysore	22.1	61	20
12. Madras	29.2	56	14
India (excluding Burma, Kashmir and Baluchistan)	33.1	79	10

[1]South-west monsoon

SOURCE: S. Banerji, "Meteorology in Indian Agriculture," *Agricultural Situation in India* issued by Economic and Statistical Adviser to the Government of India, Ministry of Food and Agriculture, August, 1961.

on labour, land and cattle, offers very little safeguard against natural hazards to individual decision-makers in Indian agriculture.

Our contention that the objective of raising family income may get modified by the considerations of maintaining the fertility of soil may be objected to on the ground that it is essentially a long-term consideration. But this long-term consideration has important bearing on the choice of crops, on the double cropping of land, etc. in the short period. That the individual producers in Indian agriculture are well aware of this necessity to maintain soil fertility over time is revealed in the intricate patterns of the rotation of crops, the widely prevalent practices of using cow-dung, river silt, oilcakes, and saltpetre as manures, and the importance attached to legume crops in the indigenous husbandry practices.[16]

It might be pointed out against our assumption that the agricultural producers in India are often observed to use cow-dung as domestic fuel rather than on fields. Dr. Voelcker, in his report on the improvement of Indian agriculture to the Government of India in 1893, categorically rejected the prevailing opinion among the officials that the "wastage" of cow-dung is due to the ignorance of the cultivators on the basis of his own analysis of the Indian soil structure and his personal observations of the ryots' behaviour in different parts of the country.[17] Dr. Voelcker advocated a policy of preserving forests in rural areas and the creation of firewood reserves in regions where harm was already done under population pressure. Dr. Walter-Leather, the Imperial Chemist to the Government of India, while agreeing with Dr. Voelcker's findings had more caustic remarks to make.

> The general consensus of opinion by those few who have really studied Indian agriculture, has been that the cultivator would not burn his dung if he was not obliged to do so. . . . On the other hand, it has been said by others that it is the ryots' ignorance which occasions the practice, and that you may give him as much fuel as you like, but he will still burn his cattle dung. . . . The same people will also tell you that you cannot improve on the ryot's agricultural practices, that he knows his business best![18]

The acceptance by individual producers of such new changes which become feasible and of the utility of which they are convinced, also lends support to the above contention. As early as 1896, Dr. Walter-Leather pointed out in his report to the Government

[16]Cf. Howard's observation in connection with the wheat producers in India, i.e. "all the methods adopted, rotation, fallowing and the growing of leguminous crops were designed for one end—to maintain the fertility of the soil." See. L.E. Howard, *Sir Albert Howard in India*, Faber and Faber Ltd., London, 1953, p. 58.

[17]Dr. Voelcker, "Improvements of Indian Agriculture," London 1893, p. 103.

[18]Dr. J. Walter-Leather, *Final Report to the Government of India* (dated 1st November, 1897), the Government of India, p. 50.

of India this aspect of the behaviour pattern of the agricultural producers in India. To quote from Dr. Walter-Leather,

> the immediate outcome of these experiments (i.e. demonstrating to the producers the manurial properties of sewage water at Meerat Experimental Farm) is that people have commenced to set up "Dhenklis" (i.e. indigenous water-raising appliances based on lever principle) to raise this "gunda nulla" water on their fields although they had expressed the opinion previously that it will burn their crops.[19]

Compared to the advanced husbandry practices based on new capital inputs such as chemical fertiliser and new knowledge of the structure of the soil, the indigenous practices in India are, however, of very little significance in maintaining soil fertility over a period of time. It has been pointed out by Nagel[20] that the southern cotton planters in the United States before the Civil War were unacquainted with the laws of modern soil chemistry, and mistakenly believed that the use of animal manure would preserve indefinitely the fertility of the cotton plantations. The same is true of the agricultural producers in India today, though there is a gap of a century between the two!

The choice variables by operating on which an individual decision-maker can maximise his family income are determined by the state of knowledge and the level of the economic development of society. For instance, since the ingredients of scientific cultivation, from fertiliser to tractor, are produced in the non-agricultural sector, the adoption of such innovations in agriculture depends on what is happening there. In fact, the above interdependence between the agricultural sector and the rest of the economy is a distinguishing characteristic of a scientifically developed agriculture as opposed to a backward one based on "traditions" and "habits." In the former, the individual decision-maker in agriculture is an important buyer of industrial outputs as his own inputs—fixed

[19] *Third Report of the Agricultural Chemist to the Government of India*, Dehra Dun, April 1896, p. 8.

[20] Cf. Ernest Nagel, *The Structure of Science Problems in the Logic of Scientific Explanation*, Routledge and Kegan Paul, London 1961, p. 476.

or variable; in the latter, the individual decision-maker produces his own inputs—manures, seeds and even sometimes implements and tools at home.

Moreover, the production of new capital inputs embodying scientific agricultural innovations in a country is dependent on the growth of agricultural engineering, chemistry, biology and such technological and experimental sciences within a country and an individual decision-maker has very little power to affect the growth of such factors individually. Guided by experiences and observations, he can select a legume crop to be grown on his land; to find out whether a particular crop suits his land and rainfall structure, (see Chapter IV) to decide, through trial and error, on the manurial properties of easily accessible things, e.g. cow-dung, river silt and oilcakes (see Chapter III) but the new horizons discovered by modern technology and experimental sciences are beyond his own limited capabilities[21] (See Chapter V).

Hence, we have incorporated the state of knowledge and the level of economic development of society in our list of constraints. This conforms to the general pattern of economic studies which analyse profit maximisation behaviour of business firms on the assumption of a given state of technique.

The inclusion of a constraint arising from the property relations in land may be questioned in view of the recent land reform legislations by the State. It appears, however, that such restrictions, arising in practice, e.g. from "feudal" types of land structure are still present in the continuation of share-cropping with absentee owners, tenancy without the security of tenure and of rent, land management by servants "attached" to families, in some parts of the country.[22]

Given the hypothesis that the agricultural producers in India maximise a function of family income as modified by two specific factors and subject to a set of constraints (discussed above),

[21]The Royal Commission on Indian Agriculture (1928) concluded; "In a country with such a long history, little surprise need be felt that a system of tillage based on experience should have reached a state beyond which further progress was bound to wait scientific discovery," *The Royal Commission on Indian Agriculture Report*, 1928, p. 14.

[22]See Daniel and Alice Thorner, *Land and Labour in India*, Asia Publishing House, 1962.

we have concentrated on the following aspects. First, we have tried to evaluate the existing husbandry practices, the choice of the cropping pattern, and adaptation to new capital inputs in terms of producers' attempt to raise their family income subject to the environmental constraints specified below (Chapters III, IV, V). Thus, we have tried to find out whether our hypothesis is not con-tracted in advance by experience that has already been observed.[23]

Secondly, we have built up optimum farm plans incorporating a new capital input, namely, chemical fertilisers for a group of family farms. Further, we have investigated parametrically into the optimum farm programmes to find out how the utilisation of land and of other resources changes on individual farms as the size of the landholding increases (Chapters VI and VII).

In the first instance, the exercises in the optimum farm-planning bear on such problems as the use of "surplus labour" in agriculture, capital needs and technical change, the profile of land use on individual family farms, etc. At a more ambitious level, these optimum farm-plans might be used as the basis of further field investigations to find out "constraints" which might inhibit producers concerned from adopting the new farm-plans.

For instance, if the logic of the farm-plans contained in Chapters VI and VII is found to be unacceptable to individual decision-makers, the hypothesis in its present form will be refuted, necessitating further investigations, e.g. in the following directions.

First, we ourselves[24] have to decide whether the inconsistency between our hypothesis and the actual experiences arises from the weaknesses in the conjectural statement itself or in the choice of the particular assumptions.

Secondly, in the case of inconsistency arising between our hypothesis and the actual experiences, we may then have to formu-

[23]A possibility which has been well emphasised by Friedman. See Milton Friedman, *Essays in Positive Economics*, the University of Chicago Press, 1953, p. 12.

[24]As has been pointed out by Braithwaite, "what experience can tell us, is that there is something wrong somewhere in the system, but we can make our choice as to which part of the system we consider to be at fault." R.B. Braithwaite, *Scientific Explanation: A Study of the Function of Theory, Probability, and Law in Science*, Cambridge University Press, 1953, p. 19.

late an alternative system in order to suggest a new hypothesis consistent with observations.

Before we proceed to discuss our hypothesis and the implications drawn from it in greater detail, it might be useful to examine a few alternative hypotheses concerning the behaviour pattern of agricultural producers in India. Although studies on Indian agriculture, in general, claim "objectivity" in terms of being a collection of data, as opposed to any deductive system or "theory," such empirical studies are invariably based on implicit hypotheses. Popper[25] has emphasised that the "collection of data" is bound to be selective, influenced by an interest, a definite task, and a point of view.

Many studies on Indian agriculture assume, implicitly or otherwise, that the determining variables in agricultural development are "sociological" rather than economic in nature.

For the sake of convenience in exposition, we have classified such sociological factors into two broad groups: (1) subjective factors, namely, religious attitudes or "values," and (2) objective factors, namely, social institutions, although these are clearly interrelated. It is suggested that such factors inhibit the possibilities of agricultural development in India. Specifically, it is claimed that, (*a*) agricultural producers in India do not have any economic motivation and are incapable of behaving rationally within the given environmental constraints, (*b*) agricultural producers in India are incapable of adapting to changes in environment, i.e. they are change resistant.

Sociologists differ among themselves in their emphasis on one or the other of the implications mentioned above.[26]

[25]Cf. K. R. Popper, *The Logic of Scientific Discovery*, Hutchinson, London, 1959, esp. Chapter 1.

[26]Hoselitz, for example, rejects the notion of the absence of economic rationality while emphasising the importance of the resistance to change. "The chief differences between Asian peasant producers and the Western stereotypycally "rational" businessman lies not in their basically different attitude towards resource allocation but rather in their different acceptance of technical change. In other words, the "traditionalism" of the Asian peasants consists mainly in his reluctance to adapt new products and new methods of production, rather than in his non-rational attitude towards resource allocation." Richard D. Lambert and Bert F. Hoselitz, *The Role of Savings and Wealth in Southern Asia and the West*, UNESCO, 1963, p. 421.

So far as the first implication is concerned, we have already pointed out that it is contradicted by the actual behaviour pattern of the agricultural producers in India.

As regards the second argument, it is doubtful whether there is any functional relationship between religious values and resistance to change. The agriculturists, who have so far adopted scientific techniques in cultivation, profess Christianity, Buddhism or Shintoism (e.g. Japan) and Islam (e.g. Asian Republics in the U.S.S.R.). On the other hand, those who are still carrying on a traditional system of cultivation, in turn, belong to Christianity (e.g. Latin America), Islam, Hinduism and Buddhism.

The attempt to put Hinduism on a differential plane from other religions, in the sense that it is particularly inhibiting to the spread of scientific cultivation in agriculture, can be objected to on the following grounds: first, as already mentioned[27] the notion that Hinduism is essentially "other worldly" is hardly sustained by expert studies on Indian philosophy, thought and logic. Secondly, the nature and content of Hinduism had changed over a period of time at different stages of development in India. Thirdly, the observed experiences of the last few decades show that the agricultural producers in India did adopt changes when such changes were feasible (refer to Chapter V).

As to the relationship between social institutions and the adaptability to change, the sociological factors appear to have more relevance to the issue at stake The following considerations lend support to the above contention:

(*a*) Historical parallelism. In all the countries with old agricultural systems, the rapid spread of scientific cultivation has been associated with far-reaching changes in social institutions. For instance, fundamental institutional changes embodied in enclosures of land may have contributed to a higher rate of growth in British agriculture, compared to that in France in the nineteenth century. Similarly, the Meiji reforms embodied in the abolition of feudal burdens on the agriculturists, e.g. rice stipends to Samurais and feudal land relationships, e.g. the establishment of individual title to land ownership, are likely to have contributed to a higher rate of

[27]See, references cited in foot-note No. 5.

growth in the Japanese agriculture compared to that in China in the late nineteenth century.

(*b*) In India, changes in agriculture have so far been sporadic and slow. It may be assumed that efforts at spreading technological changes all over the country may face resistance from the social institutions favouring the status quo. In a country such as England, the enclosure, by its very nature, broke up the status quo since it involved a natural selection of the go-ahead improving land owners and an expulsion of the multitudes of small farmers and artisans so long surviving on land.[28] Where such an abrupt change in the status quo is not feasible, efforts at economic changes need to be supplemented by deliberate efforts at social and educational changes.[29]

The question, then, is to identify which social institutions and usages are likely to inhibit the spread of scientific cultivation in the short period; the policy implications will vary accordingly.

(*c*) The spread of scientific cultivation in agriculture in India, in a broader context, is simply a means to achieve some objectives, e.g. new ideas, new cultural values and greater material comforts, rather than an end in itself. For this, again, conscious social efforts at removing outmoded usages and institutions will contribute to a speedy fulfilment of the above objectives.

In conclusion, three general comments on the sociological studies of Indian agriculture are called for.

First, the sociological explanations as they are formulated in the literature at present are extremely difficult to refute since the categories involved are non-measurable and largely non-observable as well. Secondly, such sociological explanations often confuse "static" with "dynamic" analysis. For instance, the generally accepted practice is to compare an "ideal type" of pre-industrial economy with a modern industrial society and pose the basic "traits" of the modern society as the "preconditions" of development and change, while in fact many of such "traits" (e.g. religious

[28]See G. Slater, *The English Peasantry and the Enclosure of Common Fields*, 1907 ; Also R.E. Prothero, *English Farming, Past and Present*, Longman & Co., London, 1912.

[29]This, perhaps, explains Japan's pioneering role in introducing compulsory primary education.

values, social attitudes) may well be regarded as "responses" to change.[30] Lastly, such sociological studies tend to concentrate exclusively on religion, social customs and institutions while ignoring political factors which are also likely to affect changes in the environment. If the basic question is how an individual, in a traditional society, can contribute to changes in his environment through collective decisions and actions, it appears that the crucial role is played by a "coalition" of different forces, impulses and motives which act effectively for change.[31]

To quote from Rostow's summing up: forces which have played crucial roles in changing other traditional societies were "[thus] in Germany, the coalition of Junkers and the Western men of commerce and industry; in Japan, the Samurai and the grain merchants; in post-1861 Russia, the commercial middle class and the more enterprising civil servants and soldiers."[32]

A weaker form of sociological hypothesis on which studies in Indian agriculture are sometimes based is in terms of "peasant psychology," viz. that "peasant psychology" is resistant to change and innovations. Any explanation of the continuation of the "peasant agriculture," i.e. a system of cultivation based on "experiences" and "traditions" as opposed to scientific innovations and having low labour and land productivity,[33] in terms of "peasant psychology" appears to be tautological; Indian peasants remain peasants because of their peasant psychology. If so, it is hard to find any plausible explanation why scientific cultivation has spread in France or Japan, for examples.

[30]See an illuminating discussion of the point by Professor A. Gerschenkron in *Economic Backwardness in Historical Perspective.* Cambridge, Mass., Belknap Press, 1962.

[31]Cf. W. W. Rostow, *The Stages of Economic Growth.—A Non-Communist Manifesto,* Cambridge University Press, 1960.

[32]Rostow, *ibid.,* pp. 28-9.

[33]Since a low ceiling exists in such societies on the possible rise in productivity of land and labour, agricultural producers produce mainly for self-consumption, specially when the availability of land is restricted. When a producer is an owner of large holding, he sells a large proportion of his produce in the market. See Dharam Narain, *Distribution of Marketed Surplus of Agricultural Produce by Size Level of Holding in India, 1950-51,* Asia Publishing House, 1961.

In India, the peasant producers have shown over time a positive preference for higher income and an inclination to face greater market uncertainties if there is a likelihood of greater income. This is revealed in their positive response to changes in price-ratios of competing crop or in the spread of cash crops associated with greater price uncertainties than food crops.[34] Such observed experiences contradict the notion of any peculiar peasant psychology as such.

[34]The prices of cash crops such as cotton and jute, followed international market situations and the fluctuations in their prices differed from those of food crops depending on local demand and supply. See B. P. Dutia, *Economics Aspects of Production and Marketing of Cotton in India*, unpublished Ph. D. thesis, London School of Economics, London. 1956.

CHAPTER II

RISK ELEMENTS IN INDIAN AGRICULTURE

THE PLAN of the present chapter is as follows: In the first section, we discuss some of the sources of uncertainty present in the farm-planning environment in India. In the second section, we discuss some of the economic phenomena, usually explained in purely non-economic terms, in the context of uncertainty in the farm-planning environment. In the last section, we sum up our conclusions.

THE SITUATIONS OF UNCERTAINTY IN INDIAN AGRICULTURE

In taking decisions on what to produce and how to produce, a decision-maker in Indian agriculture has to take into account four types of factors.

(*i*) *Realization of Objectives depending on an individual's own decision.* Objectives may be more than one; they may also be conflicting. One aspect of decision-making is to determine the full opportunity cost of one decision in terms of foregone alternatives. For example, the decision whether or not to invest family savings in a threshing machine may represent a choice between raising family income and maintaining liquidity at a desirable level at a particular point of time. Let objectives be denoted by the vector z.

(*ii*) *Controlled variables referring to things about which the decision-maker has freedom of choice*, e.g. how many hours to work, what type of bullocks to buy, whether to rent or to buy land, etc. Let controlled variables be denoted by the vector x.

(*iii*) *Uncontrolled variables referring to things which affect the individual's own situation, but over which he has no direct control*, e.g. how demand for cotton will fluctuate, whether there will be good rainfall, whether the government will introduce price support or control policies, etc. Let uncontrolled variables be denoted by the vector y.

(*iv*) *A structure or system linking the above variables together.* This structure or system will incorporate the relative influences of the variables and the possible effects of decisions taken. $z=f(x, y)$, taken as a formal expression of the structure, implies that the degree to which objectives will be attained depends on the level of both the controlled and uncontrolled variables.

An individual may have complete knowledge of the structure and perfect ability to predict the values of uncontrolled variables. His optimal decision would then be simply to choose values of controlled variables which will maximize his objective or a weighted average of objectives. This is a situation of certainty where the decision-maker knows values of all the parameters involved in the structure $z = f(x, y) = f(x)$ given y. That such a situation in the context of Indian agriculture is often unrealistic is generally recognised. An individual decision-maker seldom has perfect knowledge about the relationships among the variables; he generally has a limited power to predict the values of uncontrolled variables. In cases of such complex structures where z is no longer a single valued function of x, decision-making becomes difficult. In reality, there may be varied stages in between a situation of perfectly single-valued expectations and one of complete uncertainty. For example, the individual may have perfect knowledge about the structure with only imperfect power to predict uncontrolled variables.[1]

[1]Since uncertainty implies the lack of perfect knowledge of a determinate situation, decision-making under such circumstances is often regarded as necessarily irrational or random.

Such an identification of rationality with "perfect knowledge," however, is a needlessly perfectionist attitude, for there may be different levels of rationality. As Hart has observed, "actions may be rational in the light of irrational plans; plans may be rational in the light of irrational estimates, and estimates may be rational in the light of irrationally selected evidence. Hence, we are free to assume rationality at some levels without assuming it at others." Further, as is pointed out by Arrow, men are forced by the struggle for economic survival under uncertainty to set some criteria of action.

See A. G. Hart, "Anticipations, Uncertainty, and Dynamic Planning," *Studies in Business Administration*, Chicago, 1951, 11, p. 5; also, E.W. Barankin, "Toward an Objectivistic Theory of Probability," *Proceedings of the Third Berkeley Symposium*, ed., J. Neyman, California University Press, 8, 1956; also, K. J. Arrow, "Alternative Approaches to the Theory of Choice in Risk Taking Situations," *Econometrica*, Vol. XIX, October, 1951.

The situations of uncertainty present in the farm-planning environment in India can thus be classified into three broad groups.

I. Limited knowledge of the states of nature (i.e. uncontrolled variables, denoted earlier as y) which includes, in turn:

(*a*) Uncertainty arising from the role played by natural inputs.

(*b*) Uncertainty arising from variations in labour requirement due to yield variations.

(*c*) Uncertainty arising from market price fluctuations.

II. New technological possibilities when first considered by the producers (i.e. a change in the very nature of the choice variables controlled by the individual decision-maker, denoted earlier as x). Such changes reopen the question of determining the structure $z = f(x, y)$, even if the states of nature remain unchanged, since the state of knowledge regarding controlled variables has changed.

III. Changes affecting the social framework in which an individual operates, e.g. land reforms.

I. (*a*) Uncertainty Arising from the Role Played by Natural Inputs

Three factors linking the yield of crop with the natural inputs are:

(*i*) Climatic conditions, e.g. light, heat and rainfall affect both the selection and the yield of crops in a particular region. For example, a region normally having 70″ or more rainfall is likely to be rice-producing, and one having 30″ or more a wheat-producing area. A 40″ rainfall in a rice tract may turn out to be ruinous for rice although it is more than sufficient for wheat. Again, rice and wheat require different time patterns in the distribution of rainfall over a season. Thus, large deviations in the distribution of rainfall affect crop yields. Again, a year of "normal" rainfall may turn out to be ruinous to crops in a particular region depending on the timing of the availability of water.

(*ii*) The strength of the effects of variations in the natural inputs depends on such features of the soil as the depth, drainage or permeability and aeration. For example, in soils with the defective drainage or permeability, more than normal rainfall easily breeds diseases (e.g. stemrust or pgraminis of wheat) and pests; crops in soil area with a hard pan near the surface succumb more easily to weather variations than crops in a better soil structure.

(*iii*) The degree of the resistance of crops to weather variability depends also on the internal soil factors or plant food substances which can be grouped into (*a*) principal items, e.g. nitrogen, phosphate, potash and (*b*) trace elements, e.g. boron, iron, manganese. Since crops normally consume more of the elements of (*iii*) (*a*) in large doses than of (*iii*) (*b*), deficiencies in those substances easily occur; if not rectified, such deficiencies not only lower yields but also reduce the resistance of crops to seasonal variations and attacks of pests and diseases. Since elements belonging to (*iii*) (*a*) are interrelated, a chronic shortage in any one of the key elements by changing the chemical and the nutritive balance of the soil structure increases the intensity of natural hazards in agriculture —nitrogen is the obvious example in Indian agriculture.[2]

The overwhelming importance attached to the role of natural inputs in Indian agriculture arises from two factors: (*a*) The nature of the monsoon rainfall which is untimely and variable within one particular season and also from season to season. (*b*) The existing husbandry practices are relatively less effective in improving soil structure or in rectifying large deficiencies in plant foods affecting land continuously cultivated for a long time, than the husbandry practices incorporating technological innovations such as chemical fertilisers.

That the individual peasant-producer in India is aware of the uncertainty arising from natural inputs is based on the following considerations:

(*i*) Agriculture as an occupation has a continuous history of many thousand years in India, enough experiences have been

[2]In different situations, the organic structure of soils may be adversely affected by too much use of nitrogenous fertilisers. This emphasizes the importance of incorporating the modern soil science into the existing husbandry practices.

accumulated to enable an individual producer to take natural hazards into account.

This consideration is strengthened by the fact that agriculture in India is based on family farming, accumulated experiences being handed over through successive generations.

(*ii*) The agriculturists in India, in the course of time, evolved cultural practices, some of which have proved in the light of the modern scientific knowledge to be of value as barriers to nature's caprices, e.g. mixed culture with a legume crop. That many of these practices have remained unchanged so long as to appear "traditional" or "habits" may be due to the fact that the basic situation of uncertainty has not changed over time.

(*iii*) The agriculturists in India, often, try to approach the problem in a probabilistic[3] way as is suggested by their persistent (if crude) attempts at forecasting rainfall from observations on correlated natural phenomena or from past experiences of the succession of good and bad harvests.[4]

I (*b*) Uncertainty Arising from Variations in Labour Requirement

The variability in labour requirement arises from variations in natural inputs which result in variable yield per acre of land. Such variations in labour requirement are specially important for a producer in India, since agricultural operations are mainly dependent on manual labour—cattle being practically the only form of fixed capital. For example, if the crop is below normal, the labour required in harvesting is less and a part of the labour force on the farm is idle; if the crop is more than normal, a part of it may get

[3]Probability is, in the present context, measured by the degree of confidence that an individual has in the occurrence of a particular event or in the truth of a stated proposition. See I. J. Good, *Probability and the Weighing of Evidence*, London, 1950.

[4]I was told, in the course of a conversation with a share-cropper in Bengal, how he had been expecting a good harvest since the two previous harvests were poor and they had had rain in the month of Jyaistha (e.g. *Khanas*' maxims are still widely known to cultivators). In fact, if everyday language is taken as the expression of man's thought processes, then that of the peasant-producers in India is simply permeated with the probabilistic notions, e.g. widely prevalent "slokas" dealing with husbandry practices and states of nature.

lost in the absence of a reserve stock of labour to draw on at crtitical times.

Uncertainty concerning labour requirement may arise also from the supply side, i.e. members of the family are incapacitated by diseases at peak demand times. For example, specially heavy rainfall in damp regions or the growing waterlogging in many irrigated tracts often leads to the outbreak of fevers specially in times when family labour is required in fields during sowing time.

That, this element of uncertainty is taken into account by agricultural producers in India, is suggested on the following grounds: (*i*) The intricate pattern of "labour obligations" and the widely prevalent practice of annually contracted labour indicates the producer's awareness of the problem as a relevant factor in their calculations. (*ii*) If the members of the family are released to go and work in non-agricultural jobs, they are normally released, subject to the agreement that they should come back to help in harvest operations, when such help is required.

I. (*c*) Uncertainty Arising from Market Price Fluctuations

Expectations of changes in the relative price ratios of competing crops create a situation of uncertainty concerning the allocation of resources among them for an individual producer.

That the agriculturists in India are considered to take it into account as a relevant factor is justified on the following grounds. (*i*) The high short-period acreage elasticity of competing crops to the variations in price ratios can be taken as sufficiently consistent with the reality.[5]

(*ii*) The study of the long-term trends clearly reveals a long-term shift of resources in favour of the cash crops enjoying a long-term shift in prices.

II. Uncertainty Arising from Technical Changes

The new varieties of plants or animals, fertilisers and improved seeds, new ways of controlling pests, diseases and weeds, new methods of irrigation and the preparation of land, new tools and machines—issues which are currently being discussed in Indian

[5]Cf. Chapter I, p. 30.

agriculture—represent "unique" experience for the innovating producers. A little elaboration of the meaning of the term "unique" is needed. Since subjective probabilities can be "before the fact," an individual decision-maker can make his own guesses at the consequence of such events although they are "unique" to him at a point of time.[6] Whether the act of guessing is based on "partial information," "opinions," "prejudice," or "rumours," an individual decision-maker must have some implicit probabilistic estimate about consequences before he can act. "Acts" about the consequences of which an individual decision-maker knows "nothing" are simply not included in his "choice" variables. Thus, producers in Indian agriculture are likely to have different degrees of confidence in the possible good effects of chemical fertilisers on crops, their belief being based on "rumours" or "vague" information concerning what has happened on experimental farms. Thus information regarding a new innovation—both as regards amount and quality[7]—becomes crucial to the decision process of the individuals.[8]

It must be emphasised that whether available information concerning a "unique" event in economic life is regarded as adequate and reliable depends both on the objective nature of information itself and on the subjective capacity of the decision-maker. The ability of the individual to arrive at a prior probability estimate consistent with information available is determined both by what is called "intuition" and by his training in logical thought.[9] It

[6]In the present context, the Knightian dichotomy between "risk" and "uncertainty" loses much of its significance. See F. H. Knight, *Risk, Uncertainty, and Profit*, Cambridge University Press, 1921.

[7]See D. Ellsberg, "Risk, Ambiguity, and the Savage Axioms," *Quarterly Journal of Economics*, 1961.

[8]That "information" plays a crucial role in decision-making process in agriculture is evident in the role played by the "free" supply of information by the State in the technological transformation of American agriculture. See Earl O. Heady, *Agricultural Policy Under Economic Development*, Iowa State University Press, U.S.A., 1962.

[9]However, the role of "learning" in the decision-making process is still to be worked out fully. See P. Suppes, "Some Problems in the Foundations of Subjective Probability", *Information & Decision Process*, ed. R.F. Machol, Purdue University, 1960.

might explain the crucial role played by the rapid spread of education, the elaborate systems of group discussions and training in the rapid technological transformation of agriculture in Japan, the U.S.S.R., etc. Thus, technical changes affect the set of "choice" variables under an individual producer's own command.

Three observations on technical changes are called for:

(*i*) Technical changes play a crucial role in the total decision-making environment, since the sources of uncertainty already discussed are correlated with technical changes. For example, the adoption of the irrigated cultivation or the new principles of "dry" cultivation, new methods of pest, disease, weed control, the use of better tools and machines, greatly reduce the significance of yield variability due to nature's caprices, I (*a*), and variability in labour requirement and its availability, I (*b*). Once adopted, however, technical changes create their own risk situations, e.g. the possibility of the breakdown of the machines or of price fluctuations of new inputs, or of chemical fertilisers not being associated with water-supply.

(*ii*) Technical changes while reducing the importance of I (*a*) (natural factors) and I (*b*) (fluctuations in labour requirements), may increase the importance of uncertainties arising from demand fluctuations for agricultural products (i.e. I (*c*) as discussed above). This is so because the increased productivity, resulting from improved techniques, creates greater surplus output to be disposed through sales in the market. Thus, the role of self-consumption in farm-planning is reduced and the producers become market-oriented.

(*iii*) Technical changes represent new additions to the choice variables of the producer and thereby open up altogether new production possibilities which must be distinguished from efforts at achieving greater efficiency with the known choice variables.

III. Uncertainty Arising from Legal Status of Producers

Lastly, uncertainties in decision-making situations arise when there is confusion as regards the legal status of producers in relation to land. Such uncertainties may arise from, (*a*) the characteristic features of the new land reform measures undertaken by the state, or from (*b*) those of the existing system.

As regards new measures, such uncertainties may arise, (*i*) when land reform measures are protracted over a long-time horizon; (*ii*) when the objectives of land reform measures are self-contradictory; (*iii*) when legal provisions are changed repeatedly within a short period, and (*iv*) when there is a gap between legal provisions and their actual implementation.

In Indian agriculture, such environmental uncertainties are arising today from all the above factors as discussed below:

(*i*) The talk of land reform became serious as early as 1937 when the first elected governments in some states, e.g. the U.P., introduced new measures. Since 1950 onwards, land reform laws are being passed continuously in different states.

(*ii*) The objectives of land reforms professed in India are often inconsistent; for example, the legal right of the landowners to resume land for "self-cultivation" by ejecting tenants is inconsistent with the laws protecting tenancy;[10] the objective of the creation of "owner cultivators" is inconsistent with the legal sanction to share-cropping, tenancy and the very definition of "self-cultivation" as allowing cultivation and management through hired manager or agent, as in Rajasthan, and through "hired men" who may be paid in crop shares and supervised by the occupant, or by his wife or by children or by such relations as prescribed, as in Punjab.

(*iii*) The definition of "self-cultivation" for which land can be resumed by ejecting tenants varies from state to state (e.g. Bombay and Rajasthan) and may change also within a state over time. For example, a proposal in the Second Plan that personal labour should be a necessary condition of resuming land for "self-cultivation" by ejecting tenants has so long neither been officially discarded nor incorporated in legal provisions. Similarly, the

[10]For example, during four years 1951-55, the number of protected tenants and the area held by them declined by 57 per cent and 59 per cent respectively, in the state of Hyderabad. Part of the decline was due to land being consolidated by owner cultivators.

maximum limit to which land can be resumed varies from state to state[11] and also often changes within the same state (e.g. Punjab).

(*iv*) The gap between legal provisions and their actual implementation occurs in two respects: (*a*) that the laws protecting the interests of tenants have remained in statute books without any significant practical effect has been the recent conclusion reached by the Committee of the Panel of Land Reforms;[12] (*b*) since the provision of the ejectment of tenants is not associated with any provision for alternative occupation, landowners sometimes face tension and trouble in implementing legal provision. As regards the existing land tenure system, uncertainties arise from the insecurity of tenancy where tenancy is still present.

Our discussion of the uncertain situations in the farm-planning environment may be summed up as follows:

First, uncertainties arising from nature's caprices have so long played the predominant role and they are repeated. Therefore, agriculturists in India may be expected to have approximated the "optimum strategy," given the technological constraints of the economy. Once such "strategy" is reached, an individual producer is expected to adopt it repeatedly over time, giving a uniformity of motion to the system. In repeated games against nature, the "optimum strategy," if found, will be repeated by the individual concerned. Unless the expected states of nature change basically or the strategies open to the decision-maker change fundamentally, no variation can be expected in the "optimum strategy."

Secondly, once the "optimum strategy" with given choice variables is reached, any further progress of the system becomes dependent on the creation of new choice variables. The adoption

[11]The maximum limit to which land can be resumed in Punjab is 30 standard acres, not exceeding 60 ordinary acres (one standard acre means an acre of land yielding between 810-900 lbs of wheat) and 50 standard acres for displaced persons with no allowance for the size of the family; in Andhra, it is at 4-1/2 times a family holding (which is taken as 6 acres of wet and 72 acres of dry land) and for every member, in excess of five, one additional family holding with no outside limit.

[12]*The Report of the Committee of the Panel of Land Reforms*, Government of India, 1959.

of an agricultural policy by the state, oriented towards new technological changes, may be taken as the collective recognition of an individual producer's needs in the face of uncertain situations.

Thirdly, a situation of uncertainty should not be confused with a situation of change. For example, swiftly carried out land reforms may change the status quo but need not create uncertainties, if new features are precisely defined and known to the individual decision-maker. A prolonged period of land reforms with undefined and often contradictory objectives creates uncertainties for an individual decision-maker.

Economic Phenomena tied up with the Existence of Uncertain Situations in Indian Agriculture

Some of the characteristic features of the agricultural system in India appear to be tied up with the situations of uncertainty present in the farm-planning environment as discussed before. So long, such characteristic features have been sought to be explained in terms of purely sociological variables.

Following Arrow's[13] classification, we have grouped such phenomena into three classes : (*i*) Those which by their very nature are concerned with non-certain situations. (*ii*) Those which are not related to non-certain situations directly but nevertheless have no other convincing explanation. (*iii*) Those whose relation to non-certain situations is more remote and disputable.

(I) (*a*) The Payment of Water Rate

One phenomenon to which irrigation authorities in India refer to is the different behaviour pattern of the agricultural producers as regards the payment of water rate in different parts of the country. Although irrigation water is as effective in Bengal as in the Punjab in preventing crop failures in the event of a failure of the monsoon rainfall, the peasants in Bengal resist the payment of the water rate by "waiting" for the rainfall as long as they can (often too long), while peasants in the Punjab willingly

[13]See K. J. Arrow, "Alternative Approaches to the Theory of Choice in Risk Taking Situations," *Econometrica*, 1951, Vol. XIX.

pay water rate every season without any marked preference for "waiting." This observed behaviour can thus be reduced to a behaviour model:

Event. The payment of water rate to use irrigation facilities by peasants.

Let I = the payment of water rate
and II = not I
Again Let Si=good rainfall

and $\hat{Si}$=not Si
The pay-off matrix is:

	Si	$\hat{Si}$
I	$\alpha - x$	$\gamma - x$
II	α	$-\beta$

(x=the water rate)

The explanation of the pay-off matrix: we assume that the crop return is the same in the case of good rain whether irrigation water is used or not, i.e. α. Then $\alpha - x$ is the pay-off when I Si obtains. γ is the gross return, $\gamma - x$ is the pay-off when there is bad rain but irrigation water is used. $(-\beta)$ represents the loss when there is bad rain and no irrigation water is used.

According to the maximisation principle, the choice of the peasant between I and II will depend on whether EI (i.e. the expected utility from I) is greater or less than EII (i.e. the expected utility from II).

Let p be the degree of beliefs of the peasant in the occurrence Si and (1-p) in $\hat{Si}$.

Then, EI $\gtrless$ EII, according as, $p(\alpha - x) + (1-p)(\gamma - x) \gtrless p\alpha + (1-p)(-\beta)$

or $p \lesseqgtr 1 - \dfrac{x}{\gamma+\beta}$

Therefore, the actual choice will depend on the value of p, $\left(1 - \dfrac{x}{\gamma+\beta}\right)$ being given.

The accumulated meteorological data show that the probability of having a good rainfall is higher in Bengal than in Punjab.

(*b*) *Holding of Gold Stocks*

Agricultural producers in India show a strong tendency to withhold resources from use for a contingency which may or may not arise, e.g. in the form of gold stocks. From the standpoint of the individual producer, a high liquidity preference offers safeguard against sudden eventualities.

(*c*) *Multi-Cropping*

Since yields of different crops are not perfectly correlated, the diversification of product can reduce risk though it cannot eliminate it. Multi-cropping practice based on different crops, which are susceptible to weather variation in different ways, is a direct example of agricultural producers acting to mitigate risk.

(*d*) *Resistance against Devoting Total Landholding to Cash Crops*

Since food crops enjoy certainty as regards demand (via producer's self-consumption) diversion of resources to cash crops implies an increase in the zone of ignorance and risk through the operation of market forces.

(II) (*a*) *Long-term Labour Contracts*

In a world of perfect certainty, long-term labour contracts would have no significance: in a risk situation, it is a method of mitigating the undesirable effects of variations in labour requirements.

(*b*) *The Debtor-Borrower Relationship*

On the debtor's side, the need for credit often arises from fluctuations in yields, i.e. to meet the consumption requirements of the family and to pay land revenue or rent. The urgency of such loans explains these factors such as the willingness to pay a much higher rate of interest than the legally fixed rate, to enter into special obligations (e.g. to hand over surplus produce to the money-lenders).[14] On the creditor's side, the willingness to provide such loans, often on personal credit, may arise from speculative interests.

[14]See the Royal Commission on Indian Agriculture, *Report*, 1928; also, Government of India, *Proceedings of the First Meeting of the Crops and Soils Wing of the Board of Agriculture and Animal Husbandry*, 1936; also, *The Report of the Rural Credit Survey*, 1954.

(III) (*i.e. phenomena whose relation to uncertain situations is disputable.*)

(*a*) *The Trader-Producer Relationship*

The continued presence of village sales (the handing over of surplus produce by actual producers to village dealers who may be rural money-lenders, large producers, or a combination of both) may partly arise from (*b*) the debtor-borrower relationship and partly from the urge of producers to shift market uncertainties to traders as specialists.

(*b*) *The Spread of Tenancy*

Since the direct cultivation of large plots of land implies dependence on non-family labour and since such dependence is sharply, intensified in the absence of technological possibilities of using labour-saving devices, large landowners can mitigate risks arising from the uncertainty of labour supply by transferring land to tenants who supply family labour and share the effects of yield variations.

(*c*) *The tendency to combine non-agricultural occupations with agricultural operations:*

This may be ascribed to the desire to mitigate risks, returns from agricultural and non-agricultural occupations being non-correlated. Such tendency to combine agricultural occupation with non-agricultural ones is often visible not only among small agriculturists, but also among large landowners in India.

If it is accepted that considerations of income variability play an important role in the agricultural producer's economic decisions, the problem in the present context is how to incorporate such considerations in the formulation of the basic economic objective. For instance, we might assume that the producers keep the considerations of income variability as a boundary condition in maximising the objective function, viz. family income. Alternatively, it might be supposed that the producers seek to maximise a composite function incorporating both the expected income and the income variability.[15]

[15] We shall follow this second approach in our later discussion; see Chapter IV.

CONCLUSIONS

We can sum up the conclusions of the present discussion as follows.

First, while there is no generally accepted criterion of choice among the alternative approaches to decision-making under uncertainty, it may be postulated that the agricultural producers in India take into account both the expected income and income variability in their economic choices and actions.

Secondly, the degree of consistency in the choice pattern of individual decision-makers is affected by factors such as, (*i*) the nature and the reliability of the information as regards the states of nature, (*ii*) the ability of the individual to assimilate new information, (*iii*) the initial stock of capital.[16]

Thirdly, a consistent choice pattern with the given choice variables, and the states of nature may be vitiated by factors which breed insecurity and indecisiveness, e.g. socio-legal practices which may try to force a specific type of behaviour pattern on the individual independent of the latter's own preference.[17]

Lastly, in the specific situation in India, the technological innovations have the following implications:

(*a*) Technical changes, by pushing out the production possibility curve of a farm create new opportunities for higher return to the decision-maker.

(*b*) Technical changes, by changing the number of choice variables to be considered by an individual decision-maker, may change his optimum position, given the states of nature.

(*c*) Technical innovations may reduce the importance of the role of natural forces in agriculture. For instance, irrigation facilities and the modern system of dry cultivation reduce the importance of the natural supply of moisture, i.e. rainfall in cultivation.

[16]For instance, the willingness of a decision-maker to incur expenditure on a new capital input is likely to be influenced by the size of the expenditure involved in relation to his own capital stock (abstracting here from the possibility of borrowing, its cost and its effect on the initial capital stock of the producers). See L. J. Savage and others, *The Foundations of Statistical Inference: A Discussion*, Methuen, London, 1962, p. 11.

[17]For instance, the legal prohibition of cow-slaughter might be regarded as an attempt to impose a particular code of conduct on individuals, irrespective of their own preference.

The use of chemical fertilisers, together with adequate water-supply, improves the soil structure. A healthy plant is more resistant to natural hazards than a weak one due to insufficient nutrition. Improved seeds may not only raise the yield but also be disease- and drought-resistant. Improved tools and implements enable the farmer to finish his harvests quickly and more efficiently before the weather breaks. In short, technical changes reduce the role of nature in agriculture through the use of varied innovations affecting soil fertility, moisture, structure, the timings of operations, etc. The need to play for safety, faced with natural hazards, is thereby necessarily reduced.

Since the intensity of the grip of nature has an important bearing on the behaviour pattern of the decision-makers, the spread of technical changes in an old agricultural system may considerably influence their behaviour patterns. The fact that agriculture, based on scientific knowledge and innovations, is not a sport of the blind forces of nature but of human ingenuity and experimentation, may make all the difference to those who follow agriculture as an occupation.

Sir Albert Howard, on the basis of his 18 years' experience with Indian peasants, observes:

> While the average production showed no change, the seasonal variations in yields were considerable. India was found to be a country of climatic extremes not only as regards rainfall, but also with reference to temperatures, to floods and high winds. Except in certain favoured localities, the annual crop was always at the mercy of a variety of circumstances quite beyond the control of the cultivator. It is not surprising, therefore, to find him conservative in outlook, and to discover that his leading idea is to play for safety.[18]

[18]See G. L. E. Howard, *Sir Albert Howard in India*, Faber & Faber Ltd., London, 1953, p. 16.

CHAPTER III

SOME EXISTING HUSBANDRY PRACTICES

IN THIS chapter, we investigate some existing husbandry practices in India to find out how far our basic hypothesis is consistent with them.[1]

In general, the existing husbandry practices can be explained in terms of the agricultural producer's concern with raising family income without sacrificing long-term interests.

For instance, the principles of the rotation of crops are widely known and widely practised in India. Some typical examples of crop rotations prevalent in different parts of the country are as follows:[2]

(*a*) *Two years rotation*

First Year	*Second Year*
Rice and pulses	Jowar or maize
Wheat and gram	Cotton or jowar and groundnut or jowar alone

(*b*) *Three years rotation*

First Year	*Second Year*	*Third Year*
Wheat	Maize	Sugar cane
Sugar cane	Wheat	Cotton
Wheat	Wheat	Toria
Rice	Sugar cane (plant)	Sugar cane (ratoon)

[1]This review has drawn heavily on: (*a*) The technical reports left by research workers at the Pusa Agricultural Research Institute (set up by the Governor-General of India, Lord Curzon in 1904) and at Provincial Experimental Farms. (*b*) Personal reports: Dr. Voelcker and Dr. J. Walter-Leather (1887-1897), Sir Alexander Buck (1911), Sir Albert Howard (1905-1923), Sir John Russell, W. Burns, A. B. Stewart (1935-1944). (*c*) Reports of Revenue Officers contained in the annual publication of the Government of India entitled *Papers and Accounts*. (*d*) Reports of the (i) Famine Commissions of 1875, 1893 and 1945; (ii) The Royal Commission on Indian Agriculture 1928.

[2]See *Indian Crop Calendar*, Ministry of Food and Agriculture, 1956.

Given the variable nature of the rainfall, the agriculturists in India could not rigidly stick to a fixed rotational pattern of crops. Revenue Officers,[3] in course of time, have left many observations on the degree of flexibility shown by the producers in the choice of particular crops in a particular year, within the broad rotational constraints. Thus, the producers will vary their cropping pattern to meet the exigencies of abnormal rainfall conditions. The instances of producers sowing a second crop, when the first major crop determined by rotational considerations fails, are common in India.

For example, the crop statistics of Central Provinces for the year 1899-1900 (a year of severe drought) ending on the 31st May showed that the area under rice declined by 8 per cent, that under wheat by 34 per cent, compared to that of the year 1898-99, while the area under jowar increased by 15 per cent (specially in rice-growing districts), that under *tur* by 16 per cent and *til* by 48 per cent. The first two crops are more sensitive to rainfall variations than the others which are coarse low-valued crops but are more drought-resistant.

Another traditional feature of the indigenous husbandry practices is "crop mixtures," i.e. the practice of growing a major crop with a minor one. The latter is often a legume crop, which is specially noted for the nitrogen content of its leaves and roots. The advantages of the method are: First, the minor crop has properties of fixing nitrogen in the soil. Secondly, the minor crop often gives protection to the major crop by covering it against untimely or heavy rainfall. Thirdly, the minor crop, while supplying extra plant nutrients to the major crop, does not compete with it for soil moisture. The second crop draws plant food and moisture from different levels of subsoils.

On the basis of his specific investigation of the cultural practices of wheat producers in India, Sir Albert Howard concluded:[4] "It will be seen that in all the important wheat-growing districts of

[3]See *Papers and Accounts*, Government of India, 1901-02.

[4]See Sir Albert Howard, *Imperial Economic Botanist*, "Wheat in India, its Production, Varieties and Improvement," 1909, p. 33. See also F.A.O., 1953, "Legumes in Agriculture".

India, except the irrigated districts of the Punjab, the practice of growing leguminous crops on wheat lands is universal." In the irrigated districts of Punjab, Howard found that the presence of leguminous weeds, which thrive as a bottom growth in the wheat fields, enabled agricultural producers to have larger yield of wheat due to single cropping without sacrificing soil fertility over time. Such weeds ripen their seeds and dry up by the early part of April before wheat is cut and thus give no trouble at harvest time. Examples of some crop mixtures on wheat land are : wheat/gram, wheat/barley, wheat/mustard, wheat/peas, wheat/linseed, wheat/rapeseed.

The indigenous forms of manure in India are many. Dr. Voelcker[5] emphasized the knowledge of the ordinary agriculturists regarding the manurial properties of cow-dung, oilcakes, saltpetre, etc. Dr. Voelcker noted further that even when the agriculturists had prejudices against the use of night-soil[6] in a crude form on land, they were fully conscious of its manurial properties. Both Dr. Voelcker and Dr. Walter-Leather, the second Imperial Chemist to the Government of India, referred to the growing use of cow-dung as fuel under the pressure of receding green belts from villages and to the growing exports of oilseeds, and saltpetre under the attraction of the rapidly rising prices in export markets. Dr. Walter-Leather found agriculturists using earth containing nitrate when

[5]Dr. Voelcker of the Woburn Experimental Farm in England was invited by the Government of India to carry out an enquiry into the agricultural system in India and to give a report on its future possibilities. Dr. Voelcker spent five years in India visiting different parts of the country, and his report submitted in 1893 formed the basis of later actions taken by the public authorities, e.g. the setting up of the Pusa Experimental Farm in 1905.

[6]Dr. Voelcker noted the presence of prejudice among the farmers against handling night-soil directly while they were otherwise enthusiastic about the manurial properties of it. Dr. Voelcker also referred to the fact that the use of night-soil was spreading in areas where local municipal authorities had taken up the task of supplying it to farmers. Many of his successors also reached a similar conclusion. See, for example, Sir A. Howard and Y. Wad, *The Waste Products of Agriculture: Their utilisation as Humus*, 1931, Appendix A, p. 123. Also *Third Annual Report of the Agricultural Chemist to the Government of India*, Dehra Dun, April 1896, from Dr. Walter-Leather to the Secretary of the Government of India, Revenue and Agricultural Department, p. 8.

they could not buy saltpetre at its market price. Sir Edward Buck, on the basis of his own investigation, concluded that producers in different parts are well accustomed to the use of river silt as a manure and that its use on a large scale, as in Italy, depends on public authorities since it involves large engineering tasks.[7]

The preparation of the soil and the timings of sowing and harvesting in India are influenced by soil and rainfall conditions. In *sailaba* land (i.e. land either inundated or moistened by the rise of rivers), for example, the soil would be ploughed up once or twice when dry enough and levelled with the *sohaga*, i.e. a heavy square beam drawn by one or two pairs of bullocks, partly to break clods and pulverise the soil, and partly to consolidate the surface so as to retain soil moisture as much as possible. In "barani land" (i.e. dependent on rain or well water) the soil is generally ploughed more frequently during the rainy season—sometimes as many as 14 times, so that it is aerated and the rainfall is absorbed. In the case of wheat, for instance, drilling is used on "sailaba" and ' barani" soils where it is essential that the seed should reach the moisture level, while on land irrigated by well water broadcasting prevails. Different regions have different local terms to describe the physical character of the soil,[8] to classify land according to organic matter contained in soils, to indicate different requirements with regard to ploughing, etc.[9]

The practice of using natural soil aerator, such as the use of deep-rooting plants like *arhar* (Cajanusi indiens), sweet potatoes and the principles of "edge" effects in planting crops on the edges of fields, which are often slightly raised are common, specially in

[7]See Dr. J. A. Voelcker, *The Report on the Improvement of Indian Agriculture*, Government of India, London, 1893. See *Five Reports* (including the final one) submitted by the second Imperial Chemist to the Government of India, during 1893-1897. See Sir Edward Buck, *Report on the Control and the Utilisation of Rivers and Drainage for the Fertilisation of Land and the Mitigation of Malaria* to the Government of India, 1911.

[8]The soil classifications in vogue among the producers were used to build up the whole superstructure of land revenue in the 18th century.

[9]The Department of Land Records and Agriculture, Madras, recorded a number of Telegu and Tamil proverbs which are really "advices" to farmers regarding the cultivation of bad types of alkali land. See *Alkali Land and its Reclamations* Government of Madras, 1905.

regions where silt-like alluvial soils, are particularly liable to run together on the surface after rain.

Agricultural officers in India have noted the use of different varieties of ploughs and other implements for different types of soils. Another consideration, in addition to differences in soil, in the choice of implements appears to be the necessity of maintaining soil moisture. Mr. Smith, the Director of Agriculture, Bengal, concluded: "Heavier ploughs suited to Europe and America might do more harm than good in so far as they would disturb too much the subsoil and thereby set up loss of capillary water by altering the subjacent arrangement of soil particles. The illiterate Indian rayat has found this out by experience and unless some plough is devised which will not interfere with the good retention of moisture, it will be prudent not to condemn the old-fashioned Indian harrow[10].

The preoccupation of the agricultural producers with raising their family income, while protecting against rainfall variability and soil exhaustion, is reflected in the evolution of specific cultural practices designed for the above purposes. Two particular examples of such cultural practices are discussed here.

The Rab System[11]

This system consisted in burning cow-dung, branches of trees and shrubs, straw and dried grass on the ground on which seed is to be sown and from which seedlings are afterwards transplanted. The system requires a large amount of cow-dung, green materials and manual labour.

Since 1880, a large number of field experiments have been initiated to study the technical significance of Rab which farmers

[10]See for example, *Principles of Agriculture for Bengal,* Calcutta 1911, p. 31.

[11]A short bibliography on the "Rab" process:

(a) E.C. Ozanne: *Reports of the Director of Agriculture,* Bombay Presidency, 1883-84, pp. 4 and 5; 1885-86, p. 135.

(b) J. B. Knight: *Annual Report on the experimental work of the Manji and Lonavla Agricultural station,* 1906-07, p. 35.

(c) H. Mann and others: "The Rab system of Rice Cultivation in Western India," published in the *Memoirs of the Department of Agriculture in India,* Chemical Series, printed and published for the Imperial Department of Agriculture in India, Vol. II.

continue to practise while not being able to offer any cogent explanation other than that that without this particular practice they would not get rice seedlings fit for transplanting.

The following conclusions were reached on the basis of the experiments at the Pusa Experimental Farm: (i) the efficiency of the Rab system is due both to the heating of the soil and the use of the "ash," roughly 60% of the effect being due to heat and 40% to ash; (ii) normally, the weight of seedlings obtained under the Rab treatment is from one and a half times to twice as great as without the treatment; (iii) other indigenous elements such as oilcakes are not fully effective substitutes for the Rab process. (The only substitute is possibly chemical fertiliser) and (iv) possibilities of improvements on the indigenous process were found, e.g. burning the material on a heap rather than spread over a plot etc.

THE BEASI SYSTEM[12]

In rice fields when rice seedlings are one or two months old—the length of the time depending on the variety of the seeds in use—the crop is completely ploughed up, with a small plough. This is done once or twice as determined by the thickness of the crop, and fields are kept flooded to a depth of three to four inches of water. A team of women walks behind the plough and redistributes seedlings from the thicker to the thinner spots, leaving sufficient mud attached to the roots and at the same time trampling the weeds into the soil. The careful adjustment of the water level in the field is the only precaution taken by the producers while the seedlings are taking root.

In recent years, this indigenous practice has been further improved by growing a green manure crop *daincha* (i.e. sesbania aculeata) along with seedlings.

It appears from our discussion of the husbandry practices that agricultural producers in India are concerned not only with raising their family income but also with protection against natural hazards.

[12]See *Report of the second Meeting of the Working Party on Rice Breeding,* International Rice Commission, 1951, p. 34.

Moreover, they appear to be technically conscious, within the given environmental limits as determined by the state of knowledge and the structure of society.

In their concern to raise family income, agricultural producers show their preparedness to save and to invest in known agricultural inputs, e.g. land and cattle. The positive propensity to save among the agriculturists has long been recognised in India. In a recent work, Pannikar[13] has analysed family budgets collected in three widely dispersed areas which show that the gross savings ranged from 8 per cent upward of gross income per family. The first area investigated is the Punjab in the north-west of India where the individual family budget records were collected and published by the Board of Economic Enquiry since 1932-33. The second region is the jute-producing farms in eastern India, the data of which have been collected by Mr. Hans Pilhofer.[14] The third region is the state of Hyderabad in southern India.

Two recent sample surveys for the entire country show a saving-income ratio of more than 12%, excluding direct investments. As regards the investment structure in agriculture, the All India Rural Credit Survey 1954, for example, noted that the most important forms of investment in agriculture are, (i) the purchase of land, and (ii) the purchase of livestock.[15]

Similar trends were noted in all the village surveys[16] in the past and at present. According to the Rural Credit Survey, 31.5% of credit of small farmers is meant for the purchase of

[13]See P. G. K. Pannikar, "Rural Savings in India," *Economic Development and Cultural Change*, October 1961.

[14]Hans Pilhofer, *Studies of East Indian Jute Farms*, Manuscript in the Centre for International Studies, Massachusetts Institute of Technology, quoted by Pannikar, *ibid*.

[15]See *All India Rural Credit Survey*, Vol. I, part I, p. 265.

[16](i) Dr. Harold Mann, *Land and Labour in a Deccan Village*, 1917, study No. 1. (ii) Dr. D. R. Gadgil, *A Survey of A Farm Business in Wai Taluka*, Gokhale Institute of Politics and Economics, Poona, 1940. (iii) Survey Reports published by the Asia Publishing House in 1959, *A Pilot Survey of Fourteen Villages in the U.P. and the Punjab*, or *An Economic Survey, of the Village of Shamaspur* (conducted under the Agricultural Research Section, Delhi School of Economics). Also, *Bhadkad: Social and Economic Survey of a Village* (*A Comparative Study* 1915 *and* 1955), Indian Society of Agricultural Economics.

livestock. On the basis of the National Sample Survey data (Second Round and Third Round) the total capital formation for the 60 million rural households can be estimated at Rs. 5C.34 crores during 1950-51.[17] The estimate of private construction of irrigation works was about Rs. 28.4 crores for the year 1950-51, on the basis of the number of persons employed and the value of per person engaged in the construction of irrigation works.

In a post-war village study, for example, Shah has built up the following table of assets for a taluka in the south:

TABLE 12

ACQUISITION OF PHYSICAL ASSETS IN KODINAR TALUK, 1948-53

Increase (+), Decrease (—) *(rupees per family)*

Group	*Land*	*Cattle*	*Implements*
Big[a]	+147	+203	+103
Medium[b]	+ 17	+ 77	+ 40
Small[c]	— 31	+ 36	+ 10
All farmers	+ 87	+146	+ 72

(*a*) holdings larger than family holdings
(*b*) less than one family holding but larger than half of it
(*c*) less than half of a family holding.

SOURCE: C. H. Shah, *Problems of Small Farmers; Report of an Enquiry into the problems of low income farmers in Kodinar Taluka*, Indian Society of Agricultural Economics, Bombay 1958, p. 121.

[17]The four different estimates of the total value of implements per acre give the following results:

(i) West Bengal (J. P. Bhattacharji, 1945-46) Rs. 22.00 per acre
(ii) U. P. (G. D. Agarwal) Rs. 25.4 per acre
(iii) U. P. (D. S. Chanhan) Rs. 22.0 per acre
(iv) Bombay (D. R. Gadgil, 1936-37 to 1937-38) Rs. 23.6 per acre

The total value of livestock in 1949-50 was estimated to be Rs. 2,186 crores (the data of the Poona Report of the National Sample Survey adjusted to the year concerned). The total feed cost of livestock (excluding labour cost) was estimated to be Rs. 768.2 crores for 1950-51 by the National Income Committee in its final Report in 1954.

Despite the limitations to which the above statistics of saving and investment are subject, they unmistakably show a positive propensity to save and to invest among agriculturists. It is true that the peasant producers in India often save in order to buy land,[18] cattle or gold. This does not, however, refute our proposition that the agricultural producers are motivated by economic interests in their choice of husbandry practices and in their positive propensities to save and to invest in known agricultural inputs, e.g. land and cattle.

[18]About the investment of resources on the purchase of land when such purchases represented mere transfer of ownership, Calvert observed: "This investment of the crores of rupees in land does not itself add an ear of corn to the wealth of the Provinces." *Wealth and Welfare of the Punjab*, India, p. 17. In the quinquennium ending with 1877-78, for instance, average area annually sold in the Punjab was 93,000 acres and that ending with 1892-93, 338,000 acres. The investment in the purchase of land is still one of the major forms of investment in Indian agriculture.

CHAPTER IV

RISK IN FARM PLANNING AND THE CHOICE OF THE CROPPING PATTERN

ONE IMPORTANT argument against the hypothesis of the present study, namely, that the agricultural producers in India are guided by economic interests in their actions is that they are often found to choose crops to be cultivated apparently independent of their expected net return. At best, this discrepancy is sought to be explained in terms of the producer's need to supply the family's basic food requirements. However, the relative importance of different crops on individual farm units in the state of Punjab, for instance, could hardly be explained satisfactorily in terms of the above factor. Although the sugar cane enjoys a higher price than wheat in the market, producers produce extra wheat[1] and sell it in the market rather than producing more sugar cane. Again, among the food crops wheat is preferred to rice where both could be produced, while the latter yields a higher gross return (i.e. yield multiplied by price)[2]. Similarly, among the cash crops, producers show a preference for cotton to sugar cane, although the net return per acre of cotton is less than that of sugar cane.

The above crops are substitutable in the sense that if land is taken up by sugar cane, wheat or cotton could not be produced. Similarly, if land is taken up by cotton, wheat could not be produced. In the existing system of cultivation, land is not sown, in general, with two major crops, like cotton and wheat, although one belongs to the Kharif and the other to the Rabi season.[3]

[1]Punjab is a surplus wheat-producing state. Now the surplus wheat goes to other states in India; previously, it used to be exported to the world market.

[2]In the family farming system of India, the difference between gross and net return may be neglected for the present purpose.

[3]Although wheat is a Rabi crop and rice is a Kharif crop, they are often treated as substitutes. Given the existing sources of manure, if the farmers double-crop land at all, they tend to produce a minor fodder crop following a major cash or food crop.

With the spread of the irrigation facilities over the last few decades in the Punjab, the agricultural producers have changed their cropping patterns deliberately, switching over from bajra, gram, jowar, maize, etc. to wheat,[4] cotton and sugar cane. Hence the existing cropping pattern is the result of some deliberate choices on the part of the producers;[5] nevertheless, their choices appear to be independent of the economic interests involved.

Our object here is to formulate an alternative explanation of the existing cropping pattern—an explanation which takes into account both the expected income and the income variability associated with each of the crop alternatives.

The yield statistics of different crop activities show that, although the yield of each activity is affected by natural factors, e.g. moisture, temperature, energy, etc. on an individual farm unit, yields of different crops are affected differently[6] (see Table 13).

We try to find out here the possible effects of the stochastic variations in income due to yield fluctuations on the cropping pattern of two individual farms in the state of Punjab in India. Income variability in our context, arises purely from yield fluctuations due to the variable nature of the monsoon rainfall.[7] The tool of investigation used is that of quadratic programming (parametric). The scheme of the present chapter is as follows: in the first section, we discuss the aspects of the econometric tool used. In the second section, we discuss the sources of data used. This is followed by a summary of the results of the programmes. In the

[4]Hardly 17 per cent of the producers in the Amritsar District, for instance, try to produce wheat on unirrigated land affecting 3 per cent of the total acreage.

[5]The crop statistics of the state of Punjab clearly show this difference in the cropping pattern between the irrigated and the unirrigated land. See *Agricultural Statistics of the Punjab*, 1901-02 to 1935-36. The Board of Economic Enquiry, Punjab, Publication No. 52.

[6]Although crop activities are influenced by a common factor, their yields are not perfectly correlated. Hence, a combination of crop activities A and B may yield an income associated with less variability than that of either of A or of B taken separately.

[7]We abstract from the effects of income variability arising from errors in input-output coefficients. See E. O. Heady, W. Candler, *Linear Programming Methods*, Iowa State University, 1960, Chapter 17.

last section, we discuss the economic implications of the conclusion reached here, namely, that the basic features of the existing cropping pattern on the farms concerned appear to be consistent with those of that "efficient" cropping pattern which is associated with the minimum of income variation according to our programming results.

An "efficient" crop combination mentioned above is defined in the Markowitz sense[8] of a combination such that there is no other combination which gives, (1) a higher expected return and the same variance of return, or (2) a lower variance of return and the same expected return. In a parametric solution of the quadratic models, we get a series of such "efficient" crop combinations drawing out the shape of the EV (i.e. expected income and variance of income) curve.

The model of the present study is,

Maximise, $\lambda E—V$

E=the expected return from a set of crop activities

V=the variance of return of a set of crop activities

λ=constant

X_i represents the level of crop activity i contained in a set of crop activities,

R_i represents return from crop activity i

E_i represents the expected value of R_i

$V=\Sigma_i \Sigma_j X_i X_j C_{ij}$ (C_{ij} being the co-variance between activities i and j)

$$\therefore \quad \lambda E-V=\lambda \Sigma X_i E_i - \Sigma_i \Sigma_j X_i X_j C_{ij}$$

The above quadratic maximisation function is subject to the follwing constraints:

(i) technical constraint or the non-negativity constraint, e.g. $X_i \geqslant O$, for all i from I to N (i.e. crop activities cannot be pursued in negative amount),

(ii) economic constraint of $\sum_{i=1}^{n} X_i = 1$ (the entire landholding[9] must be used).

[8]See H. M. Markowitz, *Portfolio Selection, Efficient Diversification of Investments,* New York, John Wiley and Sons, 1959.

[9]It is implicitly assumed that the level of a crop activity is measured by the proportion of the total land allocated to it.

TABLE 13

EXPECTED INCOME AND INCOME VARIABILITY ASSOCIATED WITH CROP ALTERNATIVES ON THREE INDIVIDUAL FARMS, PUNJAB

(*in rupees*)

	1 *Cotton D*		2 *Cotton A*		3 *Wheat*		4 *Gram*		5 *Maize*		6 *Wheat-Gram*		7 *Toria*		8 *Sugar Cane*		9 *Rice*	
	Expected Income	*Standard Deviation*																
	(1)	(2)	(1)	(2)	(1)	(2)	(1)	(2)	(1)	(2)	(1)	(2)	(1)	(2)	(1)	(2)	(1)	(2)
Lyallpur Farm	161.50	79.33	289,0	74.80	167.65	45.45	113.47	44.64	(b)	(b)	(c)	(e)	170.48	67.48	379.88	222.35	(o)	(c)
(Canal irrigated)			(305.36)	(67.32)														
			(a)	(a)														
Amritsar Farm	164.27	56.12	(c)	(c)	194.45	57.6	(b)	(b)	126.88	89.67	109.2	41,50	(b)	(b)	293,29	223.84	245.93	119.02
	(d)								(d)	(d)					(e)	(e)	(f)	(f)
(Well and Canal irrigated)	(169.43)	(49.89)							(148.95)	(78.44)					(315.76)	(216.43)	(257.64)	(108.73)
Rohtak Farm																		
Mean Rainfal 23″ (Well-irrigated)	160.51	57.79			205.28	50.99	62.30	56.35							426.26	109.21		

NOTE :

(a) Calculated after excluding farmer's first year's yield. (b) Omitted because of too many gaps in the time series data, (c) Not grown. (d) Calculated after excluding one year of extremely bad yield. (e) Calculated after excluding two years' reported failure, (f) Calculated after excluding one year of reported failure,

The justification of this economic constraint, the operational implication of which is that the farmers concerned use land (available) fully, lies in the presence of surplus rural labour force in the region concerned. More specifically, farmers with whom we are concerned here do not appear to have any complaint as regards "labour shortage", hindering the full use of the available land during the long period for which farm records are kept (1926-27).[10]

The return from each crop activity (R_i) is assumed to be determined by a single outside influence (all crop activities are related to this basic underlying factor) and by random factors. This underlying factor is rainfall.

Thus, $R_i = A_i + B_i I + C_i$, where A_i and B_i are parameters and C_i is a random variable with an expected value of zero and a variance of Q_i and I is the level of the outside factor or the index. The future level of I is determined by random factors:

$I = A_{n+1} + C_{n+1}$, A_{n+1} is a parameter and C_{n+1} is a random variable with an expected value of zero and a variance of Q_{n+1}. The co-variance between Ci and Cj is assumed to be zero for all values of i and j ($i \neq j$).

The exclusion of the co-variance matrices from our models has the following justification: first, we have sought to capture the interrelationships among crop activities by relating them to a common index I, as is discussed above. Secondly, the method adopted here enables us to overcome the limitations arising from such conditions as the semi-positiveness of the co-variance matrix which is present, for instance, in Wolfe's[11] approach. In our models, here, we have taken into account 6 important crop activities in each case. If, however, we want to incorporate co-variance matrices in our models and follow Wolfe's method, we have to exclude 2 or 3 important crop activities, simply to satisfy the condition of the semi-positiveness of the co-variance matrices.

[10]In the context of India's low land-labour ratio, the full utilisation of the cultivable land appears to be the accepted social objective of agricultural planning. For a detailed discussion of the approach, see W. F. Sharpe, "A Simplified Model for Portfolio Analysis" in *Management Science*, January 1963, pp. 277-93.

[11]See P. Wolfe, "A Simplified Method for Quadratic Programming," *Econometrica*, Vol. 27, pp. 382-98.

The values of the parameters Ai, Bi, Qi, for each crop activity and A_{n+i} and Q_{n+i} for the index have been found through regressions on historical data. Past yields of each activity combined with the estimated future prices are used to give a measure of the expected income. This assumes that the seasonal variations in market prices due to the variations in yields are not important from the standpoint of the individual decision-maker. Heady and Candler have argued that the effect of such "compensations" due to seasonal price fluctuations can be ignored from the standpoint of individual farmer, although such "compensations" might be significant on a national basis. In Indian agriculture, it might be argued further that since nearly 2/3 of the produce is retained by producers, such "corrections" can only play a marginal role from the standpoint of the individual decision-maker.[12]

IV : I The Value of the Parameters Used and the Sources of Historical Data

Case I : Amritsar Farm			
	Ai	*Bi*	*Qi*
1. Cotton D	231.60	—3.46	3149.45
2. Wheat	130.80	3.27	3317.76
3. Maize	275.90	—7.57	8040.71
4. Sugar cane	367.90	—3.85	50108.82
5. Wheat-gram	78.42	1.57	1722.25
6. Rice	461.90	—10.44	14165.76
7. Index: Rainfall	A_{n+i}	Q_{n+i}	
	19.49	26.63	
Case II : Lyallpur Farm			
	Ai	*Bi*	*Qi*
1. Cotton D	183.00	—1.73	6293.25
2. Cotton A	292.20	— .23	5595.04
3. Wheat	154.70	1.05	2065.70
4. Sugar cane	320.20	4.78	49439.52
5. Gram	86.20	2.16	1992.73
6. Toria	185.30	—1.21	4552.20
	A_{n+i}	Q_{n+i}	
7. Index : Rainfall	12.36	26.42	

[12]McFarquhar also accepts this argument in his study of the British arable farming. See A. M. McFarquhar, *Rational Decision Making & Risk in Farm Planning—An application of Quadratic Programming in British Arable Farming*, Journal of Agricultural Economics, Vol. XIV, No. 4, 1961.

Individual farm records collected and published annually by the Punjab Board of Economic Enquiry since 1926-27 are used for our purpose. Two particular farms[13] have been selected for the present investigation: (i) Dobruji holding in the Amritsar District (1933-34/1958-59), (ii) Farm (CHAK 248 RB) in Lyallpur District (1928-29/1942-43) in the canal colony—now belongs to W. Pakistan.

Sources of Historical Data

(i) Yield[14] per acre of individual crops as actually received by these farmers on their holdings are used here. Since the source of manure remained the same, i.e. farm-yard manure, and since no other yield-improving innovation was introduced over the time, no adjustment was needed for such capital innovation. Only one example of technological innovation taking place over the period is the introduction of American cotton on the Lyallpur Farm. This could easily be taken into account as a separate crop activity.

(ii) The rainfall data on two recording stations near these two farms provide the data for the index. The distance between Dobruji farm and the station Taru Taran was 4 miles.

(iii) The expected prices of the crops selected for investigation.

[13]These individual cases could act as "test beds" for experiments in the sense that the efficient crop combinations based on our quadratic programming models might be of regional significance, i.e. for farms with similar rainfall and soil structure.

[14]Any statistical work with yield data in Indian agriculture faces two possibilities of bias: (a) the underestimate bias arising from human fallibility. The yield statistics for revenue purposes used to be collected by village officials who themselves, as landholders, might have been benefited from the lower recording of yields, since land revenue was reassessed every 25/30 years on the basis of yield records. See *Guide to Current Official Statistics*, Government of India; (b) the overestimate bias arising from the small size of the crop-cutting trials as was the prevalent practice in India. See P. C. Mahalanobis, F.R.S., "Recent Experiments in Statistical Sampling in the Indian Statistical Institute," *The Journal of the Royal Society of Statistics*, Vol. 109, 1946, pp. 342-46.

In our case, however, yield data used to be collected by the research staff of the Department of Agriculture, Punjab, responsible for maintaining annual farm records on a few selected farms. Hence, the information contained in the farm records could be used with a good deal of confidence for the present purpose.

(iv) The crops taken into account for investigation are: (*a*) *Amritsar Farm*: (i) cotton, desi, (ii) wheat, (iii) maize, (iv) wheat-gram (v) sugar cane, (vi) rice. (*b*) *Lyallpur Farm* : (i) cotton, desi, (ii) cotton, American, (iii) wheat, (iv) sugar cane, (v) gram, (vi) toria.

In the first case, the crops taken into account cover nearly 62.15% of land available, the corresponding figure is 80% in the second case.[15] The exclusion of the minor crops which account for the rest of the land available from our models is due to the defects in the statistics available, e.g. data of yields of minor crops are often lumped together and there are frequent gaps in the time series data.

FEATURES OF THE FARMS CONCERNED

(*a*) *Amritsar Farm*: (i) Land: 16.35 acres out of the total holding of 19.58 acres (loam) was irrigated.[16] (ii) Labour supply: the farmer himself and a permanently attached adult worker with the help from casual employees for specific jobs were the sources of labour. (iii) Draft cattle consisting of 4 bullocks. (iv) Average rainfall was 22.24″.

(*b*) *Lyallpur Farm* : (i) Land : the total holding was 28 acres of canal irrigated land (loam). (ii) Labour supply consisted of the members of the family, aged 65, 60, 35, 30, 14. (iii) Draft cattle numbering 4 bullocks. (iv) Average rainfall was 13.08″.

[15]The actual land distribution under the crops concerned was as follows in 1941-42:

Lyallpur Farm

Crop	*Acres*
(i) Cotton, American	7.23
(ii) Wheat	10.85
(iii) Sugar cane	1.57
(iv) Toria	0.56
(v) Gram	2.78

[16]We have taken into account only irrigated crops, yields of which are given separately from those of the unirrigated ones.

In India, weather conditions have such a profound effect on yields that even irrigation cannot offset their effects fully. See B.M. Rao, "A Study of the Eeffects of Some Weather Factors on the Yield of Wheat in Ludhiana District, Punjab." *Indian Journal of Agricultural Economics,* July-December, 1964.

(v) Technological innovation was the introduction of American cotton, in 1935-36, i.e. immediately after the great slump in cotton prices.

IV : 2 The Results Derived by Solving the Quadratic Programmes

The particular machine routine used here in solving our models[17] (i.e. Sharpe's Diagonal Model) gives us the following information:

(i) The different crop combinations associated with each corner point regarded as efficient,—efficiency as defined earlier. By arranging these crop combinations, we can draw on the *EV* curve,[18] (the Expected Income and the Variance of Income).

(ii) Associated with each combination, we are given the estimates of the expected income and the standard deviation.

(iii) Associated with each efficient combination of crops, we have the actual distribution of land among the crop alternatives as included in each combination. The share of each crop in the total land available is given in percentage figures.

IV : 3 An assessment of the Efficient crop combination lying on the E.V. curves

A qualitative assessment of the efficient crop combinations lying on the *E V* curve and of the basic features of the crop combinations as practised on individual farms concerned shows a high degree of similarity in the basic characteristics of the existing pattern and that of the corner point which lies to the left-most side of the *E V* curve (i.e. the crop combination associated with the least income variability).

We have emphasized, in our context, the following points of similarity between the existing cropping pattern and the crop combination of corner point *V*: First, multi-cropping *vs.* the

[17]Our programming exercises have been solved on the IBM 709 computer following the machine routine prepared by W. F. Sharpe.

[18]In our parametric treatment of the problem (λ varying from ∞ to 0) the quantities of the crop activities in the efficient combinations will vary until either one drops out of the combination or another enters. The point at which such a change takes place marks a new corner point. We start with that corner point which includes that crop activity which is associated with the highest expected return ($\lambda = \infty$).

FARM No. 1—AMRITSAR FARM

Corner point of each crop combination	*Crop activities contained in a corner point*	*% of land to be cropped with each activity contained in a combination*	*Standard deviation of the expected income associated with each combination*	*Expected income associated with each combination*
			(Rs.)	*(Rs.)*
I	1. Sugar cane	100	224.73	292.86
II	1. Sugar cane	34.54	117.57	270.32
	2. Rice	65.46		
III	1. Wheat	56.19	57.69	226.96
	2. Sugar cane	12.86		
	3. Rice	30.95		
IV	1. Cotton D	26.50	41.48	203.83
	2. Wheat	49.91		
	3. Sugar cane	6.59		
	4. Rice	17.00		
V	1. Cotton D	26.88	41.09	202.99
	2. Wheat	49.64		
	3. Maize	0.46		
	4. Sugar cane	6.43		
	5. Rice	16.59		

specialisation in crops. That multi-cropping, a distinguishing feature of the existing cropping pattern, associated with lower variance of income rather than the specialisation in a single crop, is clearly revealed in our programming results. The movement along *E V* curve towards the corner point *V*, in both cases, is in fact a change from single to multi-cropping. For instance, at the corner point associated with the highest expected income as well as the highest variance of income, there is complete specialisation in a single crop in both the cases. At the corner point *V*, on the other hand, 5 out of the 6 crops allowed in our two models are included in the cropping pattern (see graphs 1 and 2).

FARM NO. 2—LYALLPUR FARM

Corner point of each crop combination	*Crop activities contained in a corner point*	*% of land to be cropped with each activity contained in a combination*	*Standard deviation of the expected income associated with each combination*	*Expected income associated with each combination*
			(Rs.)	*(Rs.)*
I	1. Sugar cane	100	223.70	379.28
II	1. Cotton A	83.54	72.48	304.16
	2. Sugar cane	16.46		
III	1. Cotton A	81.29	70.34	300.16
	2. Sugar cane	15.83		
	3. Toria	2.88		
IV	1. Cotton A	72.56	62.72	285.52
	2. Wheat	7.58		
	3. Sugar cane	13.89		
	4. Toria	5.97		
V	1. Cotton D	9.45	36.74	226.41
	2. Cotton A	37.41		
	3. Wheat	31.55		
	4. Sugar cane	6.32		
	5. Toria	15.27		

Thus, by studying the shape of the *E V* curve (i.e. expected income and variance curve) in both the cases, we reach the following conclusions: (i) By giving up specialisation, the producers concerned reduce both the expected income and the income variance. (ii) The consequent reduction in income variance is much more than the reduction in the expected income, if the multi-cropping pattern is accepted.

The reduction in the income variance and expected income (in percentage) due to the giving up of the specialisation in a single crop:

	Income variance standard deviation	*Expected income*
1. Amritsar Farm	446.93	44.20
2. Lyallpur Farm	508.87	67.5

The presence of the stochastic risk elements in the farm planning environment thus encourages the producers concerned to stick to their multi-cropping structure even when market price movements may encourage specialisation in a single crop.

Secondly, the importance of individual crops in the existing cropping pattern could be related to both the expected income and the income variability associated with each crop alternative.

A few cases of individual crops will be discussed here.

(*i*) *Sugar cane* (*gur*) : Despite the high monetary return, sugar cane draws only a small portion of a producer's total land supply in the region concerned. Traditionally, this well-established practice of the agricultural producers has been ascribed either to their "insensitiveness to market forces" or to the "higher labour content" of the crop concerned. For example, it was pointed out to the Royal Commission on Indian Agriculture, "the crop does not really pay if you pay for your labour and everything else at market rates."[19] But family farmers in India do not really pay for labour and everything else at market rates!

In our quadratic programming models, 6.32 per cent of land on the Lyallpur farm and 6.43 per cent of land on the Amritsar farm

[19]See *The Royal Commission on Indian Agriculture*, Bombay, 1928, Evidence, Vol. VIII, p. 261. It was stated in evidence before the Commission that, on average, 43 per cent of land is annually devoted to wheat (cf. 49.64 per cent for Amritsar Farm and 31.55 per cent for Lyallpur Farm in the Corner Point V) and 10 per cent of land to sugar cane in Punjab as a whole.

are given to sugar cane at the Corner Point V, even though it is associated with the highest expected income. This high income variability of sugar cane might well explain why this crop never commanded more than 10 per cent of the total land in the region concerned, i.e. in the Punjab state and also on the individual farms concerned.

(*ii*) *Producer's preference for American cotton to Desi cotton.* According to the results of the quadratic programming, the relative importance of American cotton in the Corner Point V is more than that of Desi cotton. On the Lyallpur farm, the respective shares of American cotton and Desi cotton are 37.41 per cent and 9.45 per cent of the total acreage in Corner Point V respectively.

The Lyallpur farmer started producing American cotton, an exogenously introduced crop in the region, in 1935-36. By 1942-43, the acreage under American cotton has risen to 9.65 acres, representing 24.98 per cent of the total acreage (i.e. total land supply plus land double-cropped) and 34.46 per cent of the land supply excluding land double-cropped.[20] American cotton has not been included in the crop alternatives of the Amritsar farm where it has not been introduced.

(*iii*) *The producer's preference for wheat.* Although producer's preference for wheat might be partly explained in terms of the family's food requirements, wheat appears to have an important role in crop combinations based on the quadratic programming as well. For instance, in 1937-38, Amritsar farm devoted 37.55 per cent of the total land to wheat.

(*iv*) *The role of the minor crops, e.g. rice.* Despite the attractive return per acre, rice is a minor crop in the district concerned, i.e. Amritsar; the average of the district from 1954-55 to 1956-57 was 6.40 per cent of the total acreage under rice crop. The individual farmer concerned here also devoted 6.14 per cent of his total acreage cultivated to rice in 1958-59.[21] The share of this crop in the total land of Amritsar farm shows fluctuation over

[20]In the traditional system without sufficient manure supply, producers, when double-cropping, often prefer minor crops (fodder) to any major one like cotton or wheat.

[21]Rice is being grown more in the post-war years on the Amritsar farm.

the period concerned but never exceeds the above figure, i.e. 6.14 per cent.

As against our hypothesis that the choice of individual crops is related to considerations of income variability over time, it may be argued that the actual choice of crops is determined by the availability of water rather than by the above factor. One of the farms (i.e. Lyallpur farm) taken into account is irrigated through perennial canals (canals with water throughout the year) and the other (Dobruji farm) is irrigated partly by canal and partly by well. We have gone through farm accounts of the Lyallpur farm from 1928-29 to 1942-43. In one year (1932-33), the closures of the canal, at the time of sowing, affected the acreage sown for Rabi crops; nevertheless, it did not affect materially the relative importance of the major Rabi crop, e.g. wheat in the total cropping pattern. Similarly, in the case of Amritsar farm, the Kharif crop was particularly affected by water shortage in the year 1937-38. The total acreage put to Kharif crops was reduced, but the relative importance of the crops remained more or less the same.

On the Amritsar farm, the number of times the irrigation was applied to crops on the well-irrigated land does not vary much between the major crops such as cotton (3 times — 3-1/2 times) or wheat (4 times), whereas it is much less in the case of Kharif fodder crops (1.5 times). In the case of well-irrigated land, the bullock power availability might influence the relative importance of crops, since bullock power is needed to use the Persian wheel.[22] This may explain, partly, why the major crops taken into consideration in our model account for less land on the Amritsar farm[23]

[22]On the well-irrigated land, the need for draft cattle increases due to the requirements of drawing water from below the land surface. Oil engines, which can reduce this pressure on draft cattle, have recently appeared in some villages in Punjab.

[23]Estimates of manual- and bullock-labour time required for irrigation work at critical times on the Amritsar Farm.

	Days of work*			
	June	*August*	*September*	*October*
Manual Labour	9.7	7.0	26.9	30.2
Bullock labour	4.9	3.5	13.4	15.1

*(8 hours a day)

SOURCE: Board of Economic Enquiry, Punjab, *Farm Accounts, 1952-53*, p. XXVI.

(i.e. more than 60 per cent) compared to that on the Lyallpur farm (i.e. nearly 80 per cent) where the above consideration does not operate.

Lastly, the present enquiry might be used to emphasized the relationship between income variability over time and the indigenous practice of "crop-mixtures" (either a major crop with a minor one or a major crop with a legume one).

Dr. Voelcker, for the first time (in 1895), emphasized that the traditional cultural practice, namely, crop-mixtures (i.e. to produce two crops together on the same plot of land), act as protection against crop failures. Dr. Voelcker found in experiments at the Bhadgaon farm that greater profit was obtained by inter-sowing cotton with jowar or arhar than by growing it alone.[24]

While the scientific explanation why two crops, even when none of them is probably leguminous, grown together would give protection to the major crop is still far from satisfactory, we find that the income variability as calculated from the time series data of yield is higher in the case of wheat alone than in the case of wheat-gram mixture (see Table 13). Farmers in the irrigated tracts show a general preference for single-crop cultivation to crop mixtures, as opposed to the prevalent practice on unirrigated land. In the case of Amritsar farm, however, yield data of both wheat and wheat-gram are available; the variability of income in the case of wheat is found to be Rs. 3317 (associated with an expected income of Rs. 194.45) and wheat-gram mixture is Rs. 1722.25 (associated with an expected income of Rs. 109.24).

An important question arises in connection with the present study: Could we really offer any intuitive explanation why both the producers studied here appear to be approximating the last corner point on the left-hand side of the *EV* curve which is associated with the minimum of standard deviation of income? Without more intensive

[24]See Dr. J.A. Voelcker, *The Report on the Improvement of Indian Agriculture,* Government of India, 1893, p. 234. Dr. Walter-Leather, the Second Agricultural Chemist to the Government of India, while agreeing with the basic observations of Dr. Voelcker on the importance of indigenous cultural practices the agricultural producers inIndia, was sceptical about the degree of the effectiveness of crop mixtures against yield variability. See J. Walter-Leather, *Final Report to the Government of India,* Dehra Dun, India, 1897.

enquiry into the problem in different regions, we could not reach any definite answer. However, it appears from the shape of the *EV* curve that any significant rise in the expected income is associated with a very large increase in the income variability; for instance, in the case of the Amritsar farm, a 44.20 per cent rise in the expected income is associated with 446.93 per cent increase in the income variance (measured by standard deviation), the corresponding figures for the Lyallpur farm are 67.5 per cent and 508.87 per cent. In a situation as described above, producers appear to have little opportunity to raise the family income without increasing the income variability to a very large extent, indeed!

IV : 4 Conclusions

To sum up the current discussion, we reach the following conclusions. First, the actual cropping patterns existing on the farms concerned are not inconsistent with the hypothesis that the agricultural producers in India maximise a composite function of the expected income and the income variability. The importance of the latter arises from the presence of risk elements in the farm planning environment due to the role of natural factors in Indian agriculture.[25]

Secondly, the possibilities of raising farm income in the short period by introducing greater specialisation in high-valued crops appear to be limited in the region concerned, given the relationship between the income variability and the expected income (see Table 13).

In a recent study, for example, Desai has concluded that a 20 per cent rise in farm income is possible without any technical change if the producers concerned could be persuaded to change their existing cropping patterns in the districts of Ahmednagar and Nasik in the Bombay State.[26]

[25]It must be emphasized here that the conclusions reached are subject to an important limitation, namely, the extremely small size of the sample (i.e. two farms in one region). The present investigation, however, has opened up a new direction of enquiry which, if pursued, might give us a better understanding into the actual behaviour pattern of the agricultural producers in India.

[26]D. K. Desai, *Increasing Income and Production in Indian Farming*, The Indian Society of Agricultural Economics, Bombay, 1963.

Desai's conclusions, though not applied to the region with which we are concerned, appear to be subject to the following limitations: (i) Desai's choice of the cropping pattern is based on linear programming with single-valued expectations (i.e. certainty). (ii) The region studied by Desai is specially susceptible to yield variability arising from the variability of the rainfall.[27] (iii) The major part of the land affected by Desai's cropping pattern is unirrigated on which the yield variability is much more significant than on the irrigated land with which we are concerned here.

If the argument as developed in the present section is consistent with the reality, Desai's conclusions based on an argument which ignores the presence of the stochastic risk elements as reflected in the relationship between the expected income and the income variability might turn out to be of limited significance in actual farm planning. If so, the method of raising farm income is to be sought elsewhere, e.g. improving factor productivities[28] which might increase output and income without increasing the importance of the stochastic risk elements unduly. For instance, technical changes by raising the expected income from each combination of crops associated with the given variance of income may offer a new opportunity to individual producers to raise their family income and farm productivity. Such opportunity of raising income will be practicable for individual decision-makers, given their existing risk aversion functions. Over a long time horizon, the importance of risk elements arising from natural factors may themselves be

[27]See Harold Mann, "*Rainfall and Famine*"—*A Study of Rainfall in the Bombay Deccan, 1865-1938*. The Indian Society of Agricultural Economies, Bombay, 1955. Cf. Mann's observation : "I doubt whether there are many agricultural centres in the world in which cultivators depend on the rainfall... and yet where the variability is as great as in the tract represented, at least, by the last three stations (i.e. *Ahmednagar*, Sholapur, Bijapur) under study" (p. 16.).

[28]For an individual producer in India, the opportunity of raising farm income by extending the scale of operation with the given technique is limited by the non-availability of fresh land. On farms, where unused cultivable land exists now, such opportunities are however available in the short period.

reduced through the spread of scientific cultivation based on new knowledge and new capital inputs.[29]

The significance of such technical changes, however, will largely be dependent in the short period, on their impact on the existing risk elements now present in the farm-planning environment and on their physical productivities, i.e. the capacity to raise the expected income.

[29]An interesting line of future investigation is to re-examine our quadratic models by introducing the fertilised processes of the crop activities concerned. This might throw light on the possible changes in the expected income and income variance associated with the efficient crop combinations when technical innovations are adopted. Our failure to pursue this line of investigation is due to the non-availability of time-series data of yields of crop activities based on chemical fertilisers and such innovations.

CHAPTER V

INNOVATIONS, UNCERTAINTY AND THE OPTIMUM FARM DECISIONS

THE SCHEME of the present chapter is as follows. The first section discusses some of the characteristic features of technological innovations in agriculture. In the second section, we assess technological innovations in terms of their implications for uncertainty arising from the role of natural factors in Indian agriculture. This is followed by a discussion of how producers adapted themselves to innovations which were made feasible from their standpoint through the deliberate efforts made by the public authorities over the last few decades.[1] In the last section we draw some conclusions from the present discussion, which have a bearing on further spread of innovations in Indian agriculture.

V : 1 TECHNOLOGICAL INNOVATIONS

Technological innovations can be defined as the use of new inputs and knowledge leading to an outward shift of the production possibility envelope. Given the existing state of human knowledge, the more important technological possibilities in Indian agriculture include machines—large and small, chemical fertilisers, hybrid seeds, insecticides, modern drainage systems, irrigation methods, principles of dry cultivation, the new methods of soil preservation and cleaner and more scientific ways of handling the night-soil.[2]

[1]See Sir John Russell, *Report on the Work of the Imperial Council of Agricultural Research in Applying Science to Crop Production in India*, Government of India, 1937.

[2]The fact, that the scientific cultivation based on technological innovations spread, for the first time, in countries with a practical need to save labour more than land, has often led people to equate technical change with the use of tractors — an idea depicted in the following statement, e.g. "in Asia technical change is a question of adapting to, and not adopting, technological innovations," the reason being the population pressures on land. See *Proceedings of the International Conference of Agricultural Economists, 9th Conference*, 1955, p. 94.

Although a single major input is seldom substituted for another single input,[3] technological innovations in agriculture can be classified, broadly, as (i) relatively labour-saving (e.g. tractors and threshing machines), or (ii) relatively land-saving (e.g. fertiliser, hybrid seeds).[4]

Technological innovations in agriculture are often complementary[5] to each other; the use of one innovation creates the need for the use of another. When a group of innovations is used at a time, each becomes more effective in raising output than when used individually. Thus, chemical fertiliser and hybrid seeds are more effective in raising output when used as a composite dose than when used as single inputs. The effectiveness of the above composite dose is, in turn, dependent on the availability of sufficient water, and manual labour for hand weeding or chemical weedicides. The bumper crop following the use of the chemical fertiliser, hybrid seeds, water and weedicides may require better implements of crop cutting than the traditional sickle of the Indian farmer and better systems of harvesting than the traditional ones. Similarly, deeply ploughed land requires more regular water supply and a greater availability of plant nutrients than when ploughing is shallow. Hence, innovations are more effective, in the short period, in pushing out the production possibility envelope when used in a 'packet' than when used individually. The spread of irrigation facilities in the Punjab without a simultaneous spread of new drainage systems

[3]See E.O. Heady, "Marginal Rates of Substitution Between Technology, Land and Labour," *Journal of Farm Economics*, February, 1963.

[4]Technical innovations in the manufacturing sector are generally categorised into labour-saving, capital-saving or neutral. See R.M. Solow, "Technical Change and the Aggregate Production Function," *Review of Economics and Statistics*, August, 1957. It might be interesting here to refer to the experiences in Japan where the emphasis was primarily on raising the yield per acre of land vis-a-vis those in a group of countries (U.S.A., Australia) where the emphasis was to raise output per head of worker in the early stages of economic growth. Farm labour force per hectare of arable land in Japan is 160 times that of Australia, 70 times that of the U.S.A., and 7 times that of Italy. See H. T. Oshima, "A Strategy for Asian Development," *Economic Development and Cultural Change*, April 1962.

[5]Cf. *Agriculture: Science and Technology for Development, Report on the United Nations Conference on the Application of Science and Technology for the Benefit of the Less Developed Areas*, Vol. III, United Nations, New York, 1963.

is an example of the failure to recognise complementarity among investments. A similar example is the spread of improved seeds of rice in the southern states without a simultaneous arrangement for the supply of more plant nutrients.

Complementarity among innovations creates the practical possibilities of having increasing returns to their use. Thus, the use of improved seed and chemical fertiliser, no doubt, requires more water, but the water requirement is less than proportionate to the rise in yield. For example, in the Punjab, an application of 40 lb of nitrogen per acre in the form of sulphate of ammonia saves 6″ depth of irrigation water in the case of sugar cane.[6] The need for a "package deal" in Indian agriculture arises largely from the above complementarity among innovations.

To conclude, any possibility of introducing innovations in agriculture requires, in the first instance, the growth of scientific knowledge in the country concerned. No practice is universally beneficial; in fact, what is good for one system of cultivation may be positively harmful for another with a different soil-climatic nexus. Thus, there is no short-cut in the selection of innovations suitable for a particular country. Given the state of human knowledge, deliberate efforts have to be made to find innovations suited to a given context. This search for knowledge and innovations is a never-ending process. As Russell reminds us, " ... each generation seeks its own proofs and soon discovers that the supposed facts and principles of its predecessors are only partly true."[7] A continuous search for new knowledge forms the very basis of any possibility of technical change in Indian agriculture. Its requirements have to be decided in accordance with the natural conditions of the country and the needs of the individuals concerned.

At a point of time, there is, however, a great deal in existing scientific knowledge that is sufficiently well established to be applicable on farm units. In the next section, we discuss how far

[6]See K. N. Raj, *Some Aspects of the Economics of Bhakra Nangal Project*, 1960, Asia Publishing House, London.

[7]John Russell, *Report on the Work of the Imperial Council of Agricultural Research in Applying Science to Crop Production in India*, Government of India, 1937, p. 4.

scientific knowledge has been related to the requirements of Indian agriculture and how far practical steps have been taken to bring new knowledge and new ingredients of science to the village level.

V : 2 Innovations and the Role of Natural Inputs in Agriculture

Agriculture as an occupation is subject to uncertainties arising from the role played by such natural factors as heat and sunshine, moisture and rainfall, plant nutrients and soil properties. While some degree of exposure to natural forces is inevitable, it is argued below that technological innovations can greatly reduce the importance of uncertainties arising from the above factors in Indian agriculture. The main sources of uncertainty arising from natural hazards in India are: (i) climatic factors, (ii) plant nutrients, (iii) soil conditions such as depth, drainage and permeability, (iv) plant diseases.

The Control of Moisture

(i) The problem arises from the situation where the customary source of moisture is variable and often insufficient due to the nature of the monsoon rainfalls.

The results of the variable and often low moisture supply are: (*a*) low yield in regions with low rainfall, (*b*) great variations in yield due to variations in moisture supply, and (*c*) complete crop failure due either to floods or to droughts.

The solution to the problem of low and variable moisture supply lies in the spread of irrigation facilities.

Irrigation facilities act as an "insurance"[8] against natural hazards by preventing serious setbacks due to the lack of moisture. By allowing land to be moistened when it is otherwise too dry to plough, irrigation facilities allow increased flexibility of the timing of operations and obviate the need to keep land fallow due to the shortage of water. Moreover, the spread of irrigation facilities makes it feasible to introduce further innovations such as fertilisers

[8]See O.W. Israelsen, *Irrigation, Principles and Practices,* Second edition, John Wiley and Sons, 1958.

and hybrid seeds which, otherwise, may be excluded from the producers' choice variables. Greater human control over the supply of moisture would make it possible to stabilise yield of crops at a high level.

In India, yields of crops are both higher and steadier[9] on irrigated land as compared to unirrigated land in the same region with the same crops. For instance, the average yield of wheat is 1,055 lb on irrigated land and 565 lb on unirrigated land in Amritsar District in Punjab; the corresponding figures for the Ferozepur District are 1,000 lb and 485 lb respectively.

Quinquennial ending	*Irrigated*		*Unirrigated*	
	Matured%	*Failed%*	*Matured%*	*Failed%*
1905-06	91.1	8.9	71.6	28.4
1910-11	90.5	9.5	70.8	29.2
1915-16	93.2	6.8	74.8	25.2
1920-21	92.6	7.4	74.1	25.9
1925-26	94.0	6.0	78.6	21.4
1930-31	92.7	7.3	71.3	28.7
1935-36	93.1	6.9	73.4	26.6

SOURCE : *The Agricultural Statistics of the Punjab* (1901-02 *to* 1935-36), Board of Economic Enquiry, Punjab, 1937, p. 25.

Other solutions to the problem of the lack of moisture lie in the adoption of better varieties of drought-resistent species. That this is a practical possibility is suggested by recent developments in the science of plant life and by the favourable experimental results secured in India. Again, the adoption of dry cultivation based on contour bunding, trenching, gully-plugging and afforestation can reduce yield variability, raise average yield per acre and reduce soil erosion,[10] where irrigation facilities cannot be

[9]For Quinquennial Average Maturity and Failures of Crops on Irrigated and Unirrigated Land in the Punjab, see the table above.

[10]Sir Albert Howard concluded that the yield of cotton per acre can be doubled on the black soils by a balanced combination of agronomy genetics and soil science, e.g. fresh supply of organic matter and manures improving surface, drainage, new methods of eradicating *Kans* (a type of perennial grasses), seed selection, etc.

created.[11] The adoption of modern land development and soil conservation measures,[12] flood control systems and schemes of afforestation can similarly lessen the frequency and the intensity of floods and droughts.

The Control of Plant Nutrients

The problem arises from the fact that the cultivation of land disturbs the equilibrium of natural nutrients in soil and its nitrogen content declines with crop production. In the early stages of cultivation of virgin land, crop yields tend to be high due to the abundance of nutrients derived from the high organic nitrogen reservoir. Unless deliberate efforts are made to stabilise the nitrogen content of the soil by applying manures, the loss of natural nutrients may create a situation where land may go out of cultivation. This actually happened in many countries and in experimental farms where plots were kept unfertilised.

In many regions in India where continuous cultivation is very old, the level of soil fertility has been stabilised by the farmers resorting to the use of manures such as cow-dung, oilcakes, river silt, saltpetre, green manures, to the mixed cultivation of major crops with legume crops (a widely prevalent traditional practice in India), to practices of crop rotations, etc. These indigenous practices have stabilised the soil nitrogen content over time, but it has stabilised it at a low level. Burns stated, "Indian soils are at a stage in which, on the whole, there is neither increased nor diminished production. Judging from the results of over 5,000 manurial experiments in India and the variability in the yields of non-manure plots, it is probable that in most parts of India soil fertility is stabilised at a comparatively low level."[13]

[11] A scheme for improving eroded land based on contour bunding, trenching, gully-plugging, afforestation, etc. undertaken in Bijapur, Sholapur and Poona Districts in 1943 (covering 125,000 acres in Bijapur), showed a doubling of the average yield of *Rabi* crops.

[12] See *Regional Meeting on Land Utilisation in Tropical Areas of Asia and the Far East,* Ceylon, 1951, *Minutes of the Sessions,* F.A.O., p. 82 ; *Soil Erosion by Wind and Measures for its Control on Agricultural Lands,* F.A.O., 1960.

[13] See W. Burns, *Technological Possibilities of Agricultural Development in India,* Government Press, Lahore, 1944, VII (Summary); also A. B. Stewart, *Report on Soil Fertility Investigations in India with Special Reference to Manuring,* Army Press, Delhi, 1947.

Factors responsible for the stability of soil fertility at a low level in India are:

(*a*) The inescapable loss due to the sheer operation of time, i.e. the long history of continuous cultivation over many centuries in the major agricultural regions in India.

(*b*) Environmental factors such as climate, rainfall. For example, soil nitrogen and carbon in India rise with increasing mean annual precipitation both in the hot coastal areas and in the cool Himalayan ranges. Again, there is a strong negative correlation between nitrogen and carbon on the one hand and temperature on the other. As many of the major agricultural regions in India are associated with both high temperature and low mean annual precipitation, environmental factors contribute to a rapid loss of soil nutrients. The importance of such environmental factors, as the temperature and the level of rainfall in the context of India, has been emphasised by Jenny and Ray Chowdhuri in their pioneering work on the Indian soil structure.[14]

(*c*) Cultural factors like exhausting soil management practices (the scarcity of wood and coal, leading to the use of cow-dung as fuel, the absence of green manure crops, the failure due to population pressures to resort to fallowing) have contributed to the loss of nitrogen and organic reserves in the soil. For example, the loss of nitrogen and organic reserves is much less on the cool Himalayan tracts with ample supply of firewood and in relatively sparsely populated areas in the Deccan than in the Indo-Gangetic Plain long associated with population pressures and receding lines of natural vegetation for more than 100 years.

A theoretical calculation, based on rates of gains and losses of nitrogen in the American system of cultivation, suggested that the near-steady state in nitrogen and carbon reserves is reached in less than a century of cultivation and that the cultivation index or N_x[15] is 26.3 per cent for a very exhaustive cropping system

[14] See H. Jenny and S. P. Ray Chowdhuri, *Effects of Climate and Cultivation on Nitrogen and Organic Matter Reserves in Indian Soil*, The Indian Council of Agricultural Research, 1960.

[15] The cultivation index is the ratio of nitrogen in the soil cultivated (Nc) to that in the corresponding virgin soil (Nv) multiplied by 100, or $Nx = \frac{100Nc}{Nv}$.

(continuous corn). The lowest array of Indian Nx values (in regions with cultivation for many centuries) is 35, 32, 31, and 28 per cent.

Prior to the widespread use of hybrids and nitrogen fertilisation, average corn (maize) yields in the central part of the U.S.A. was thus conditioned by soil nitrogen percentages.

$y=224.6N-3.3$ ($r=0.995$), y denotes corn yields (in bushels) per acre. The insertion of the nitrogen content of humid Indo-Gangetic soils ($N=0.056$—0.06 per cent) seems to predict a corn yield from 8 to 10 bushels per acre. The average corn yield in the U.P. is actually near about 600 lb/A or 11 bu/A.[16]

The low nutrient content of the soil increases the susceptibility of the plant to natural hazards (minor weather variations) and increases the possibilities of plant diseases. A low yield per acre and a great variability in yield are the results.

For example, the partial failure of the American cotton crop in the Punjab in the years 1919, 1920, 1921, 1926, 1927, 1928, 1931, and 1932 (the external symptoms being the early rendering and shedding of leaves, the premature opening of the bolls with immature seeds and the dwarfing of the plants) were found to be due to the malnutrition of the plants owing to the presence of alkali-salts in the subsoil region and the extreme nitrogen deficiency in the soil, and not to factors such as the shortage of water.[17]

The solution to the problem of the low nutritive reserves in the soil lies in the greater use of plant *nutrients* available in varied forms. One important source of plant *nutrients* lies in the chemical fertilisers, a new technological possibility. A balanced use of manures based on a combination of the indigenous sources of plant nutrients as discussed earlier along with modern innovations may improve the soil structure over time.[18]

[16] See Jenny and Ray Chowdhuri, op cit., p. 100.

[17] Department of Agriculture, Punjab, *A Summary of the More Important Results Arrived at or Indicated by the Agricultural Stations and Research Officers in the Punjab, During the Years* 1935-36 *to* 1937-38, Lahore, 1940, p. 89.

[18] See *Scientific Reports of the Indian Agricultural Research Institute*, 1961; A. B. Stewart, *Report on Soil Fertility*, 1947; also V. Ignatrieff, *Results of Experiments and Practical Experiences in the Use of Fertilisers, Manures, and Soil Amendments with Rice*, F.A.O., 1951 (section on India); Albert Howard

Jenny and Ray Chowdhuri on the basis of their study of Indian soil observe: "It is concluded that the low level of organic matter in many Indian soils is primarily caused by environment, and only secondarily, by cultural practices. Since hot climates mitigate the build-up of soil organic matter, the solution of *India's nitrogen economy* must be based on nitrogen fixation by industrial plants."[19]

It must be acknowledged here that the greater use of plant *nutrients* is effective only when used in conjunction with other complementary inputs, e.g. water. Further, in principle, the excess of plant nutrients may be as much of a problem as a shortage. In the present context, however, it is the shortage of plant nutrients that need worry us.

In all the experiments so far carried out in India, soils of all types (irrigated) have shown positive response to nutritive treatments. Recent experiments on unirrigated land, similarly, show positive response.[20]

All India 2 years (1954-55 and 1955-56) Average Response to Chemical Fertiliser when Applied to Paddy and Wheat

Fertiliser treatment (lb per acre)	*2-Year yield (lb)*	*Increase over control (lb)*	*Return to fertiliser investment (%)*
Paddy control	2,316	..	..
30 N as ammonium sulphate	2,666	350	90
30 N $30P_2O_5$	3,080	764	170
Wheat control (irrigated)	1,159	..	..
30 N as ammonium sulphate	1,382	223	83
30 N $30P_2O_5$	1,656	497	173

SELLING PRICES:

(1) Paddy — Rs. 10 per md (2) Wheat — Rs. 15 per md
(3) 30 lb N as ammonia sulphate Rs. 23 (4) 30 N 30 lb P_2O_5 Rs. 34.4
(60% super charged first year).

See, for earlier experiment results, Stewart and others.

See *Report of the Indian Council of Agricultural Research*, 1957-58, p. 14.

---

and G.L.C. Howard, *The Application of Science to Crop Production*, Oxford University Press, 1929, especially p. 27 and 30; Sir John Russell, *Report on the Work of the Imperial Council*, Delhi, 1937.

[19]H. Jenny and S.P. Ray Chowdhuri, op. cit., 1960, p. 100-01.

[20]For Summary of the Results of Fertiliser Demonstration Programmes (Joint Indo-American Scheme) during 1954-57, see the table above.

To conclude, there are large possibilities of improvement in the supply of plant nutrients in land. Till very recently, an important technical innovation almost completely bypassed Indian agriculture, i.e. chemical fertiliser. The recent improvements in its use (see later discussion) have covered only a very small portion of the cultivable land having regular and heavy rainfall or having irrigation water all the year round. Moreover, many easily available sources of plant nutrients are wasted in the existing system, e.g. cow-dung,[21] oilcakes, river silt. Efforts might be made to use them more on land. Similarly, if chemical fertiliser is in short supply, alternatives might be found in less spectacular sources, e.g. sewage water, night-soil, water haycinth, etc. Although the nitrogen content of such resources is low,[22] they are easily available within the country.

The Control of Soil Structure

Generally, the soil scientists have still a lot to discover about the characteristics of the soil structure[23] and their relevance to cultivation.

The problems which need urgently the attention of the soil scientists in Indian agriculture today are connected with the waterlogging of land, the rise in the subsoil water level in the irrigated tracts,[24] the rise in the salinity of the soil and the erosion of top soil. The above situations are making land unusable for further

[21]During the Second Five Year Plan, a programme was launched to conserve cow-dung in the community development blocks in order to raise the production of farmyard manure by 24.5 million tons by the end of the plan period; the target for the Third Five Year Plan is 70 million tons spread over 3,600 community development blocks. In the absence of information we could not make any proper assessment of the actual improvement.

[22]The nitrogen content of Ammonium Sulphate is 20.6% as against 0.5% of farmyard manure and 5 to 7.1% of oilcakes.

[23]See H. Jenny, *Factors of Soil Formation*, McGraw Hill Book Co., New York, 1941; also C.E. Kellogg, *Conflicting Doctrines about Soil*, F.A.O., 1957.

[24]See *Triennial Review of Irrigation in India*, 1918-21, Public Works Department, Government of India, *Triennial Review*, 1933-36, Government of India ; H. H. Mann, *The Economic Results and Possibilities of Irrigation*, *Indian Journal of Agricultural Economics*, June, 1958; *Proceedings of the Board of Agriculture in India* held on 9th December 1929, Calcutta, 1931; and M.B. Williams, *Essential Considerations for Irrigation Development*, F.A.O., 1951.

cultivation, and increasing the frequency of plant diseases (e.g. infestation in the waterlogged transplanted rice fields in South India) and the sensitiveness of plant life to natural hazards in different parts of the country.

Technically speaking, the solutions to the above problems lie in the proper drainage systems, underground and surface and in the scientific study of the soil, its profile, structure and content followed by modern soil treatments. Whether such steps will be economically feasible in a particular region will depend on a cost-benefit calculation of specific projects.

The Control of Diseases and Insects

The traditional cultural practices are singularly ineffective in controlling plant diseases and insects in Indian agriculture. The result of backward agricultural technique is that the low yields per unit of land often arise from plant diseases and insects.[25]

The problem of low yield arising from insects and diseases could be tackled by (*i*) the spread of modern methods of disease control through cultural practices, such as, the flooding of the fields, the cutting of stalks at ground level, etc. Plant disease is often most injurious in waterlogged transplanted rice fields and in double-cropped areas. Damage to crop varies from 3 per cent to 65 per cent in Southern India.

Cultural methods of controlling Schenobious incertellus include: (*a*) cutting of rice stalks at ground level and storing them in such a way as to prevent further larval development or the escaping of the adults, (*b*) flooding of fields — specially, immediately after harvest, (*c*) removal of infested plants and their destruction. In India, it is recommended, that stubble should be uprooted and burned in a large area before March for at least 2 seasons, (*d*) the practice of autumn or winter ploughing which kills many larvae hibernating in the stubble in the field, (*e*) sowing of early matured varieties, and (*f*) the practice of clean cultivation.

[25]In Southern India's rice fields, losses due to infestation are sometimes as high as 87 to 90 per cent of the crop (e.g. Schenobious incertellus is very damaging in some districts in South India while Rice Hispa causes severe damage to deep-water rice in Bengal and in Assam areas). See *Report of the Second Meeting on Rice Breeding*, International Rice Commission 1951 (especially Report by K. Ramiah).

To remove the already deposited eggs, mechanical methods include hand collection of egg-masses, and clipping of leaf blades during transplanting.

Similar to the above examples, other cultural and biological methods of controlling many other crop insects and crop diseases require the dissemination of scientific knowledge among the cultivators and a lot of organising ability at the village level.

(*ii*) The introduction of the improved species, e.g. in the varietal resistance tests in India, wheat variety N.P842 was found to be resistent to all the prevalent races of black rust, E4360 to all yellow rust.[26] Similarly, in the case of the flint type of maize, most popular among the producers in India, 4 types of hybrid maize have been developed which show greater resistance to diseases and pests than either the local varieties or the dent grained American hybrids; in addition, a 50 per cent rise in yield is shown at a recommended level of fertilisation over the local variety.[27]

The implications of the introduction of improved cultural practices on individual farm units have not, however, yet been empirically investigated.

V : 3 THE EXAMPLES OF INNOVATIONS INTRODUCED IN INDIAN AGRICULTURE AND PRODUCERS' ADAPTABILITY TO THEM

I. Varietal Selection

Improved varieties of crops and seeds have been associated with higher income and lower variability whenever they have been introduced into Indian agriculture. A few examples are mentioned below.

(*a*) *American cotton*[28] : Deliberate efforts were made by public authorities to spread this better variety of crop in the Punjab. The

[26]See *Scientific Reports of the Indian Agricultural Research Institute*, 1961.

[27]*Ibid.*

[28]The last decades of the 19th century witnessed the beginning of a steadily growing demand for raw cotton from the Indian and the Japanese mills and the average annual production of raw cotton (including Burma) rose from 1.6 million bales to 3.7 million bales by 1906-7. The product was, however, short staple indigenous variety. The stoppage of imports of long staple cotton during the World War I emphasised the need for its production at home. In 1921, the Indian Central Cotton Committee, with statutory power and special funds from a cess on domestic consumption and the exports of cotton was set up to facilitate the introduction of better varieties of cotton.

success of such efforts has traditionally been ascribed to the income effect of the change. To quote from the Report of the Board of Economic Enquiry, Punjab, "Prior to 1921-22, all cottons were returned under Desi (short staple, indigenous) cotton and, therefore, the area under American cotton is not traceable earlier than this year (1921-22). The conservative agriculturists are convinced that an acre of American cotton brings them more money than an acre of the indigenous variety and, consequently, the area under the former is gradually increasing."[29]

An investigation into the yield variability shows that the variance in yield of American cotton over time is lower than that in Desi cotton.[30]

(*b*) *Improved Seeds of Wheat*: In the case of wheat,[31] specially in the Punjab and in the U.P., deliberate efforts were made by agricultural departments to produce and to supply improved seeds to the farmers. The following table indicates the rapidity in the spread of improved seeds.

TABLE 14

ACREAGE UNDER IMPROVED WHEAT IN THE PUNJAB (1918-1932)

Year	8-*A*
1918-19	18
1922-23	55,000
1927-28	1,276,449
1931-32	2,549,060

SOURCE : Board of Economic Enquiry, Punjab, *Agricultural Statistics of the Punjab,* 1901-02 *to* 1935-36, ed. J.W. Thomas, 1937, p. 23.

[29]Acreage (irrigated) under American cotton rose from 401,182 in 1921-22 to 1,275,954 in 1935-36, and the acreage (unirrigated) rose from 199 in 1921-22 to 29,503, i.e. many times within a period of 15 years. See *Agricultural Statistics of the Punjab*, 1901-02 to 1935-36, Board of Economic Enquiry, J. W. Thomas, ed., 1937, p. 23.

[30]We have investigated the case of irrigated American and Desi cotton (see Chapter IV); The crop combination, with the lowest income variability under quadratic programming, allocates more land to American cotton than to the Desi variety.

[31]One of the major food grains export from India till 1920.

The total acreage under wheat (irrigated) fluctuated between 4,740,338 and 5,590,717 acres during the above period.

By the end of the decade, the *limiting factor on the spread of improved seeds became the shortage of supply*. In 1937-38, the Agricultural Department reported, "in the case of 9D — the problem is to meet demand from areas of Kangra Valley, . . . in the case of C591, released in 1934, the available supplies cannot meet demand already generated."[32] In 1961 as well, it was reported that the supply of improved seeds falls short of demand; for instance, in the case of New Pusa variety, only 61,885 kgs of seed could be supplied against a demand of 267,230 kgs.[33]

II. Nitrogenous Fertilisers[34]

The deliberate efforts to supply the information of this new input to actual producers is a post-war phenomenon in India. By 1960, the gap between the supply and the demand for nitrogenous fertilisers has widened considerably. Against a total production of nitrogenous fertilisers of 90,000 tons in 1960, the estimated minimum quantities to be required by the end of the Third Five Year Plan (i.e. 1965) is 1.25 million tons.[35]

	1955-56	1956-57	1957-58	1958-59
(1) Ammonium sulphate	93,885	120,581	84,045	90,705
(2) Ammonium sulphate nitrate	2,233	2,213	9,280	10,530
(3) Urea	2,268	1,585	7,547	14,825
(4) Calcium ammonium nitrate	1,457	nil	nil	nil

SOURCE : *Proceedings of the 13th Meeting of the Crops and Soils Wing of the Board of Agriculture and Animal Husbandry in India*, June 1960, p. 264.

[32]A Summary of the More Important Results arrived at or indicated by the Agricultural Stations and Research Officers in the Punjab, during the years 1935-36 to 1937-38, Lahore, 1940, pp. 2 and 4.

[33]See *Scientific Reports of the Indian Agricultural Research Institute* 1961, p. 22.

[34]For Consumption of Nitrogenous Fertilisers on the Basis of Actually Distributed Amount, see the table above.

[35]See *The Proceedings of the 13th Meeting of the Crops and Soils Wing*, 1960, p. 72.

On the basis of their study of individual farms, Mellor and Moorti concluded, "The problem is *once more* of physical *supply of inputs and demonstration* and not credit or low income as such."[36] Reddaway mentions "embarrassing shortages" of chemical fertilisers in the market due to the scarcity of foreign exchange resources.[37]

The use of nitrogenous fertilisers, given sufficient water, not only increases yield considerably but also reduces yield variability by improving soil conditions in the Indian situation.

III. Improved Tools and Machines

(*a*) *Heavy ploughs.* The one example of unsuccessful efforts at introducing innovations by the agricultural departments in India is the soil-inverting deep and heavy ploughs.[38] The explanation of this failure is as follows.

(*i*) The heavy ploughs were not liked by the farmers. The present investigator feels that this dislike was due to the fact that these ploughs were not suited to the soil and rainfall conditions[39] and would have increased yield variability as the recent scientific enquiries tend to show.

(*ii*) These imported ploughs were priced considerably high[40] as compared to the indigenous smaller ploughs. The demand for the latter remained strong throughout this period.

[36] *Ibid.*, p . 162.

[37] W. B. Reddaway, *The Development of the Indian Economy*, George Allen and Unwin Ltd., 1962, p. 27.

[38] Officers complained that the experimental farms were becoming show-cases of unsold implements. See *Catalogue of Implements and Machinery Found Useful for India* (1882/83-1897/98); also Sir John Russell, *The Report on the Working of the Imperial Council of Agricultural Research*, 1937, p. 1.

[39] See Sir Albert Howard's unpublished Report quoted by E. Howard in *Sir Albert Howard in India*, Faber & Faber, 1953, p. 227; F. Smith, *Principles of Agriculture for Bengal*, 1911, Calcutta, p.31; also *Essential Considerations in Mechanization of Farming*, F.A.O., 1950.

[40] The price-consciousness of the ryots led the Institute of Plant Industry at Indore to evolve a new system of eradicating *Kans* in cotton fields. First, the work with heavy, soil-inverting ploughs was found to be unsatisfactory. The employment of the tractor and the steam plough was out of the question—to quote from Sir Albert Howard—"as the use of such costly implements would for all time effectively prevent the Institute from obtaining any influence with the cultivators. When the cost of an implement approaches the purchase

(*iii*) Sometimes it has been pointed out by the officials that the imported ploughs were not suitable for being drawn by draft cattle that farmers had.

(*iv*) There was no arrangement for the repairs and servicing of these implements (imported from England and Switzerland) at village levels.

(*b*) *The improved varieties of the indigenous implements.*[41] There is much scope for improvements in the indigenous implements made and repaired by village artisans (often farmers supplying raw materials)—specially in their designs[42] (lessening the intensity of work for the man and the animal), in the raw materials used[43] and in the methods of fabrication; there is scope for introducing small improved tools and implements for specific operations, e.g. harvesting.[44]

When efforts have been made to produce better designed indigenous implements without any significant rise in prices, the producers have responded; for instance, the cases of Bose plough and all-iron plough (DRG 10) from Tanjore District, Madras.

price of a small village, the cultivator argues thus: why cultivate at all? Why not buy a village, become a zamindar and live comfortably on the rents?" By modifying an American ridging plough (from which the wings and sole have been removed) drawn by two pairs of oxen, the Institute succeeded in evolving a set of cheap implements which were sold to the cultivators as well. See A. and Gabrielle Howard, *The Application of Science to Crop Production*, Oxford, 1929, p. 30.

[41]Classes of indigenous implements in use are : (i) Ploughs—different types for dry, wet, and irrigated land, (ii) harrows, (iii) rollers, levellers, (iv) clod crushers, (v) seed drills, (vi) hoes, (vii) ridges, (viii) harvesting implements such as sickles. In general, the indigenous implements tend to be shorter-lived and lower-priced than their foreign counterparts.

[42]Assuming that the average furrow width is 5 inches and assuming that no gap is left between two furrows, the ploughman with such an indigenous tool has to walk at least about 20 miles per day to cover an acre. Improved designs can considerably reduce the strain on man. In the case of seed drills, indigenous designs cannot combine the work of seed-cum-fertiliser placement, no mechanical adjustment is allowed to vary the seed rate, not able to sow mixed crops.

[43]Indigenous harrows become easily blunt as only mild steel is used. See *Indigenous Agricultural Implements of India: An All-India Survey* (of 20 states), The Indian Council of Agricultural Research, 1960; also *The Mechanization of Small Farms in European Countries*, O.E.E.C., 1950.

[44]See *Small Farm Implements*, F.A.O., 1953.

The limiting factors appear to be, (*i*) the failure of the agricultural departments to undertake or finance research in new designs, (*ii*) the absence of special financial concessions for facilitating the production of new implements on a mass scale and at a low cost, (*iii*) the failure to create proper marketing arrangements so that the prices of the improved implements are not inflated by too many middlemen, (*iv*) the failure to disseminate new information among producers and artisans who may have to undertake repairs

IV. Irrigation Facilities

The easy adaptability of the individual producers to irrigation facilities in India is well known.

The effect of the spread of irrigated cultivation, when such facilities were created, was a rise in income[45] and a fall in income variability. Since irrigated cultivation offers greater opportunities for manoeuvring with traditional knowledge and skill, individual producers in India often utilised this opportunity to raise family income. Two such examples are cited here.

(*i*) Producers in the irrigated tracts in the Punjab, evolved the practice of growing gram in *wadh watter* (i.e. a crop sown in the moisture left by the irrigation to the previous crop, e.g. gram sown in the *wadh watter* of rice or chari), so that no separate irrigation is required for its preparatory tillage and sowing. Obviously, this practice to sow gram immediately after the harvesting of the preceding crop permits the ploughing of fields only once or twice. According to the cultivators, this does not affect the yield of the crop and a common proverb expresses their ideas on the point: *jat ki jane kah, chhola ki jane wah*, "Just as a Jat (a cultivator) does not care to follow a regular path, the gram crop does not care for ploughing."[46]

(*ii*) Sir Albert Howard found as a result of his specific investigation that in all the important wheat-growing districts in

[45] The standard yield of wheat on irrigated land is 1,037 lb in the Punjab, 1,200 lb in the U.P., while that on unirrigated land is 572 lb and 800 lb respectively. Producers in Punjab rarely cultivate wheat on unirrigated land. Since all-India yield statistics lump together yield on all types of land, irrigated or unirrigated, aggregative data conceal the effects of technical improvements.

[46] See *The Cost of Production of Gram in the Punjab*, 1954-55 *to* 1956-57, The Board of Economic Enquiry, Punjab, 1960, p. 1.

India, except the irrigated districts in the Punjab, the practice of growing leguminous crops on wheat land[47] is a universal and traditional practice. The presence of leguminous weeds such as Melilotus Indica, Melilotus Alba and Medicago Deuticular which thrived as a bottom growth in the wheat fields, however, enabled the producers in the canal colonies to give up the traditional mixed culture to single product and raise total output considerably.[48]

V. Dry Cultivation and Improvements in the Non-Irrigated Regions

Since such technological improvements are still confined to the experimental farm stage[49] and efforts have yet to be made to bring new information and new capital inputs to individual producers, instances of changes in unirrigated tracts are few. Two examples of such changes are the following.

Improved seeds used on unirrigated land	*Increase in yield* %
Maize	15
Cotton	10
Oilseeds	10

(*i*) Wheat producers themselves in the Punjab started producing an improved variety of wheat (C 591) on unirrigated land. Though C 591 was originally given out for cultivation on good average soils under irrigation, it came to be grown under all sorts of conditions; irrigated or *barani* (i.e. rain-fed), on good and poor soils, early and late, etc.

[47] Some of the common mixtures: wheat with gram, barley or mustard, wheat-peas, wheat-linseed, wheat-rapeseed (oil-seeds are specially noted for their nitrogen fixing properties). Such crop mixtures on unirrigated land have been found to provide security against crop failures while maintaining soil fertility.

[48] When wheat is produced in mixture with legume crops, the yield of wheat is low compared to its yield when cropped individually.

[49] Some experiments recently conducted by the Indian Agricultural Research Institute, under the Indo-U.S. Fertiliser Programme, show the possibilities of increased yield by using improved seeds on unirrigated land.

(*ii*) The spread of American cotton on unirrigated land: Although the introduction of American cotton was specially related to irrigated cotton land, it gradually spread over the unirrigated land as well. The proportion of unirrigated land used for cotton is, however, very low compared to irrigated land used for the same purpose.

VI. Pest, Disease Control, Improved Methods of Planting, etc.

The actual adoption of innovations in these directions are hard to find because of the following reasons:

(*i*) Agricultural research institutes have included only recently the problems of pest, disease control in their fields of enquiry on a large scale.

(*ii*) Whatever results were secured earlier are still largely confined to the experimental farms.

VII. Drainage, Afforestation, Control of Floods, Waterlogging, etc.

Since such improvements have to be conducted on a large scale involving the expenditure of large sums of capital and other resources, the undertaking of such innovations is dependent on the public authorities[50] rather than on individual farmers.

Drainage on a small scale which can be undertaken by the producer needs the spread of new knowledge among the cultivators.

[50]The schemes of land development and soil conservation measures requiring bunding, trenching and similar steps can be undertaken only on the basis of a complete watershed or a particular region; often such schemes involve the limitation of individual property rights (ref. Soil Conservation Laws of America, Australia, New Zealand, etc.) See *Soil Conservation : An International Study*, F.A.O., 1948. One example may be cited here to illustrate our point. In the area adjoining the Jumna and Chambal rivers in the U.P., for example, 3 classes of land—deeply ravined, land affected by varying degrees of active sheet and gully erosion, apparently or relatively unaffected table lands—all have to be covered by one single integrated scheme. Excessive run-off from better lands at the top of a drainage area tends to cause damage to the ravines below, and conversely, failure to stabilise ravines by protective measures tends to undo any improved practices applied to lands at the head of the ravines. See *Report to the Government of the State of U.P.*, F.A.O., 1955. The participation of public authorities is needed in securing the services of the experts, raising finance etc. However, the ultimate success of such measures depends on whether individual producers participate by contributing money or labour, and whether they are taught new principles of soil management.

Systems of surface drainage, devised at experimental farms, have not yet reached the actual producers.

V : 4 Conclusions

I. The Major Limitations on Individual Producers

(*a*) The shortage of new inputs representing innovations: The new capital inputs for the adoption of which efforts have so far been made are all in short supply; while over vast agricultural regions, efforts have yet to be made to bring information and resources to individual producers. For instance, in the case of further spread of improved seeds, the limitation is arising from the shortage of supply, the current demand of the producers cannot be met. Similarly, the current demand for chemical fertiliser is outstripping the current supply (cf. Section V : 3). The absence of real resources explains the failure of individual producers to utilise credit accommodations offered to them. In a recent supervised credit programme near Delhi,[51] loan amounts were made not subject to the maximum limit but left to be determined by the need of the cultivator and his ability to repay from increased production. The average size of short-term loans for 1957-58 and 1958-59 Rabi crops were Rs. 40 and Rs. 34 respectively, while the medium-term loans for the purchase of bullocks and buffaloes (known inputs) averaged Rs. 368. It was estimated that a short-term loan of Rs. 200 per farm could have fertilised a large portion of land. No attempt was, however, made to relate the offer of credit to the physical availability of fertiliser in the region. The credit offers remained largely unutilised.

(*b*) *The high price of new inputs since they are either being produced for the first time or are being imported.* For instance, the price of Ammonium Sulphate (subsidised) is Rs. 392.09 per ton in India and Rs. 298.64 in Japan.

(*c*) *The absence of facilities to acquire information regarding new innovations.* The use of mass media, e.g. radio, lantern lectures or disseminating information in villages, is used on a very small scale. Since most of the agricultural producers are illiterate, the above mentioned means of communication appear to have special significance in India.

[51] See *Proceedings of the 13th Meeting of the Crops and Soils Wing, 1960.*

(*d*) *The absence of facilities to acquire training in assimilating information* (*if present*) *through group discussions, organised trials, and elementary education.* The lack of opportunity as reflected in the above limitation is the inhibiting factor on the individual decision-maker in Indian agriculture — a lack of opportunity which can be ascribed to the limited growth of industries[52] and to the limited efforts of the public authorities to demonstrate innovations at the village level.

Heady[53] sums up the experiences in the U.S.A.: "Technical discovery or innovation, as much from the industries outside agriculture as from the public research organisations attached to the farming industry, has caused these capital items (i.e. representing innovations) to be known and made available." In the context of India, Mellor emphasises that the supply of unconventional inputs (e.g. fertiliser, hybrid seeds) or developmental services depends very largely on the government's initiative and programmes.[54]

II. Major Limitations on the Public Authorities

(*a*) *The shortage of foreign exchange resources*[55]. Many of the new capital inputs or at least the machines for their production at home

[52]For a discussion of the relationship between the spread of technical changes raising agricultural productivity and the growth of industries, see Deane and Cole, *British Economic Growth*, 1688-1959; *Trends and Structure*, Cambridge University Press, 1962, especially pp. 62-75; also L. H. Dupriez, ed., *Economic Progress* (Lovain), 1955, especially the contribution by Prof. Habakkuk. Beaumont and Higgs observe: "Although changes in the farmer's implements had become inevitable, the actual time of change was to some extent conditioned by contemporary industrial progress. In 1750, most tools, used by farmers, were made, either locally, by the village blacksmith or carpenter, or, in a cruder form, by the farmer himself. In 1850, the equipment of the farm was coming from the factories of industrial Britain." See Singers and others, *A History of Technology*, Oxford, 1958, IV, p. 1. For the case of America, see H.U. Faulkner, *American Economic History*, 7th edition.

[53]See E. O. Heady and Baum, ed., *Capital and Credit Needs in a Changing Agriculture*, Iowa State University Press, 1961, p. 124.

[54]J. W. Mellor, "Increasing Agricultural Production in Early Stages of Economic Development, Relationships, Problems, Prospects," *Indian Journal of Agricultural Economics*, 1962, April-June, p. 44.

[55]Given the rate of increase in population every year, we can feed new additions to population for a period of five years, for example, by importing food (minimum amount required), or we can import fertiliser every year to produce the same amount of food at home, or we can import fertiliser

have to be imported from abroad. For instance, the production of chemical fertilisers even when raw materials are available is being restricted by the shortage of foreign exchange resources. Similarly, the use of better material for the production of improved tools and implements is inhibited by the shortage of iron and steel.

(*b*) *The shortage of manpower*. Technological innovations in agriculture require the services of Agronomy, Microbiology, Agricultural Chemistry, Agricultural Engineering, Mycology, etc. The limited growth of scientific education acts as an inhibiting factor on the large-scale undertakings of agricultural research.[56]

(*c*) *The weaknesses of the administrative system.* Regarding improved seeds, for example, Nillson-Leissner concludes, "without organised and supervised production and the distribution of high quality seeds of improved varieties, the efforts of the scientists and plant breeders will be more or less in vain."[57]

machines to produce the required amount of fertiliser every year at home; in all the cases, foreign exchange resources are needed. Foreign exchange requirements either for import of food grains to feed new additions to population for a period of five years, or to import fertilisers enough to produce food at home, or to import machines to produce the same amount of fertiliser at home are given below.

	I Year	*II Year*	*III Year*	*IV Year*	*V Year*
1. Food grains (*a*)	33.6	67.2	100.8	134.4	168.0
2. Fertiliser (*b*)	10.5	21.0	31.5	42.0	52.5
3. Fertiliser plant (*c*)	16.8	16.8	16.8	16.8	16.8

SOURCE : *The Proceedings of the* 13*th Meeting of the Crops and Soils Wing of the Board of Agriculture and Animal Husbandry in India*, 1960, p. 269. See also F.A.O., *Fertilisers Manufacturing Projects—their Planning and Financing*, 1949.

[56] As Professor Solow has emphasised : "Almost always, technology is assumed to change autonomously; in purest principle there is a deliberate location of resources to the generation of technological change—one can think of this as a kind of super technology." See R. Solow, *Capital Theory and the Rate of Return*, Amsterdam, 1963, p. 39. The Second Five Year Plan allocated Rs. 14.15 crore for agricultural research in India.

[57] See G. Nillson-Leissner, "The Organisation of Seed Prodcution and Distribution Schemes," in *Production and Distribution of High Quality Seeds of Improved Varieties*, F.A.O.

The agricultural departments ascribed their ability to supply only 61,885 kgms of new Pusa wheat as against a demand for 267,230 kgms in 1961 to the limited facilities available for seed multiplication. The efforts of the scientists regarding dry cultivation and soil preservation have remained confined to experimental stages due to the absence of organisation to take them to the individual producers.

(*d*) *The lack of organisation at the village level.* The need for such organisations arises in India: (*i*) to act as links between the agricultural departments and individual producers for the supply of new inputs and for the dissemination of new knowledge; (*ii*) to supervise the use of new inputs. That the lack of such supervision has often led to the wastage of new inputs, e.g. irrigation water, seeds in India, is shown by many studies;[58] (*iii*) to provide a channel through which the individual requirements for the marketing and the credit can be related to the higher organisations. For instance, village cooperatives in Japan assist rice farmers to get their seed requirement each year in exchange of their products. In the context of India, Stewart emphasised that the "official village agricultural store may be of great assistance to the extension officer";[59] (*iv*) to give an organised expression to the opinions, views and the ideas of the individual decision-makers. So far, there has been only one-way traffic, namely, from the Extension Officer to the individual producer.

The need for such village organisations has been emphasised by men who studied the problem of technological changes in Indian agriculture.[60]

[58]In India, defective administrative systems often encourage the wastage of resources. For instance, the water-rate not estimated on the volumetric rate tends to encourage waste of water.

[59]See A. B. Stewart, *Report on Soil Fertility ... Manuring*, Army Press, Delhi, 1947, p. 49.

[60]Small producers scattered over thousands of villages create a special problem for the administrative staffs of the agricultural departments, namely, how to keep contact with the innovating producers. Because of the presence of illiteracy on an overwhelming magnitude in Indian villages, personal contact often is the only alternative. If producers are organised in village associations, this problem could be handled through them. However, the success of village organisations depends on factors such as social cohesion, prospects of concrete improvements in life, the leadership among the producers; there is no easy way out of the problem.

(e) *The size of the task.* The technological innovations in India are to cover 326,297 thousand hectares of land and to affect 62 million producers scattered over 560,000 villages.

(f) *The large variations in climatic, soil and cropping patterns.* These create the need for flexibility in the choice of technological innovations for a particular region in India. The lack of imagination and the knowledge of the locality concerned often account for suggestions for improvements which leave individuals concerned unconvinced.

III. Other Factors

For individual farmers in India, the adoption of a capital item or a new cultural practice representing innovations appears to be determined by its implications for the family income as we have argued earlier. When an innovation represents a rise in farm income, individual farmers take it into consideration. Factors determining whether such innovations are adopted, among others, appear to include the value of the product, the *price* ratio of the factors and the physical productivity of new inputs. For instance, the adoption of American cotton was helped, among other things, by the high price of the product relative to competing crops; the adoption of improved seeds and chemical fertilisers has been helped by the high physical productivity of these inputs. On the other hand, the labour-saving devices may be relatively less attractive because of the low wage rate of casual labour.[61]

For the individual farmer, a second consideration in the adoption of innovations, as discussed earlier, appears to be the implication for uncertainty present in the farm-planning environment. If an innovation tends to reduce the income variability, its adoption is easier.

When the considerations for family income and of the uncertainty arising from natural factors diverge, the actual adoption of an innovation depends on the individual producer's risk preference and the time horizon of planning; in such cases, the response of the producers can be expected to vary from case to case.

[61]Cf., E.O. Heady, "Farm Use of Capital in Relation to Technical Change and Factor Price," in *Capital and Credit Needs in a Changing Agriculture*, ed. Baum, et al., State University, Press 1961.

Our discussion so far tends to bear out the adaptability of producers in India to innovations when such opportunities were created through the active efforts of the public authorities in the past. It suggests that agricultural producers may well respond to new opportunities in future, if such opportunities are created in such a way as to be consistent with their own requirements.

The above conclusion is, however, subject to the following limitations: First, since prediction invariably involves 'induction,' we have to be specially cautious in taking any prediction about human behaviour at its face value.

Secondly, the efforts at introducing innovations have so far been slow and rather sporadic. Moreover, such efforts were largely confined to particular regions, e.g. the Punjab. When efforts are made to spread innovations rapidly and all over the country, we might find non-economic variables (e.g. social institutions, customs, the lack of education in the villages) inhibiting such efforts.

Thirdly, the spread of technical changes appears to have taken long time in already developed countries;[62] in a country such as India, great allowance for time might well be needed.

Sir John Russell emphasised this crucial role of the time element in the present context. To quote from Russell:

> When the application (i.e. of science) has been worked out, the cultivator still has to be persuaded to adopt it. This often takes a long time and the cultivator is abused as slow and conservative because he does not change his methods more rapidly. But it is frequently impossible for him to do this. His unit of time is a year; 'safety first' must always be his guiding principle, and methods known to give subsistence to himself and his family cannot possibly be discarded until the superiority and practicability of the new ones are established beyond question. Even when he is convinced of their advantage, he may still lack the means for adopting them. Further, he thinks in terms of his entire system of cultivation and an isolated process, though good

[62] See O.E.E.C., *Manures and Fertilisers : Potential Progress in Europe*, Paris, 1959.

> in itself, may not fit well in with his general management. This extension work, therefore, requires something more than education and propaganda for the new methods; it involves much adjustment of the new methods to the actual conditions and systems on the cultivator's land.[63]

The real possibility of the spread of innovations in agriculture arises, in the last instance, from the individual's concern for higher income, for a better standard of living for their families, specially in a society where the process of economic growth itself creates new wants[64] for new goods, for better quality in the old goods, new opportunities for education, for medical care; in short, the economic motivation becomes the determining factor. In the short period, however, the socio-political structure of a country might rouse these latent aspirations of the individuals and canalise them for rapid economic development, or it might frustrate such aspirations leading to chaos, or to apathy.

[63]Sir John Russell, *The Report on the Working of the Imperial Council*, 1937, p. 4. A fundamental change in a given socio-political set-up might hasten up the spread of technical change in agriculture. The setting up of the large 'enclosed' farms in the U.K. or the large 'collective farms' in the U.S.S.R. might be taken as examples of such changes.

[64]Cf. Joan Robinson, *The Accumulation of Capital*, Macmillan, London, 1957, p. 296; also, J. W. Mellor, "The Use and Productivity of Farm Family Labour in Early Stages of Agricultural Development," *Journal of Farm Economics*, August, 1963.

CHAPTER VI

NEW CAPITAL INPUTS AND OPTIMUM FARM PLANNING

IN CONNECTION with our enquiry into the rationale behind the choice of cropping patterns on two individual farms in the Punjab, we identified a positive relationship between the expected income and the income variability associated with alternative crop patterns. On the basis of the above data, we solved two Quadratic Programming models where the function to be maximised was 'E-V' where E is the expected income and V the income variance. A qualitative comparison of the existing cropping pattern with the crop combinations lying on the EV curve, i.e. the curve which gives the most efficient combinations[1] of crop activities showed that the actual cropping pattern on the farms concerned approximates a particular crop combination lying on the EV curve (cf. Chapter IV).

In the present chapter, we intend to derive new optimum farm plans for a group of farms by using the linear programming[2] technique to find out what would happen if producers behave as if they are maximising their expected income 'E' rather than 'AE-V'; in short, if they stop bothering about 'risks' involved in the farm planning environment.

The plausibility of the exercise depends on a number of crucial assumptions: First, an innovation epitomised in the form of 'chemical fertilizer' is available at a known price; individual farmers can buy fertilisers as much as they want for their farm planning. Secondly, individual farmers concerned have reliable data on the basis of which they could formulate subjective estimates of the effects of using chemical fertilisers.[3] Thirdly, farmers have

[1]See Chapter IV.

[2]See E. O. Heady, W. Candler, *Linear Programming Methods*, Iowa State University Press, 1960.

[3]The 'reliability' of data is determined by the producers' willingness to act on their basis.

an assured supply of extra water required if they use chemical fertiliser. Fourthly, farmers have an assurance from the public authorities of 'full compensation' if their income after change is less than that they could expect in the absence of a change; farmers have confidence in the assurances given.

In such circumstances, it might not be too unrealistic to assume that farmers agree to give a trial to the innovation epitomised in the chemical fertilisers.

Farm plans, built up on the above assumptions, are not strictly speaking behaviouristic models; they might rather be described as logical exercises setting forth 'normative' programmes which producers might adopt to achieve the maximisation of the family income.[4] Such plans, in our context, are expected to tell us not only what individual producers 'ought to do' but also what the public authorities ought to do to create an atmosphere where individual farmers might get a reasonable chance of raising their family income.

In the present chapter, we intend to draw farm plans on a group of farms based on the above assumptions. For this purpose, we could not use the individual farms considered earlier under the quadratic programming because we did not have input-output coefficients and other necessary data for the building up of new farm programmes.[5]

The scheme of the present chapter is as follows: In the first section we describe the models of individual farms and discuss some of the technical problems that arise in using the linear programming technique in the present context. This is followed by a section incorporating the results of the programming models. In the last section, we try to draw some conclusions assuming that individuals concerned would try out our alternative farm plans. Such implications will have an important bearing on the formulation of the public policies.

[4] See B. J. Diggs, "A Technical Ought," *Mind*, Vol. LXIX, 1960.

[5] We could not investigate the present farms earlier through the quadratic programming since we do not have time series data of yields on these farms.

VI: 1 Models

The region selected for the present investigation is the state of Punjab in North West India. The villages were selected by Daniel at random from the districts of Amritsar and Ferozepur and the holdings[6] selected by the stratified random sampling method, the sample being the representative of the holdings of different sizes as existing in the villages. Criteria for selecting particular holdings were the size of the farm (in acres), homogeneity of soil, type of farming, size of family, etc. The farm records used were obtained from the cost accounting surveys made in the above districts by the personnel of the Farm Management Research Centre and applicable most directly to farms on such lands in the districts concerned.[7]

Agricultural Features of the Region

(i) *Types of soil.* Mostly alluvial soil, deficient in nitrogen, humus and phosphoric acid; sufficient in lime and potash noted for the quick renewal of nitrogen; capable of producing a wide variety of crops including rice, wheat, sugar cane, cotton, tobacco, maize, bajra, gram, oil seeds of different varieties, fruits, e.g. banana, melons, etc. The entire areas of the districts concerned, namely, Amritsar and Ferozepur, are covered by such a soil.

(ii) *Rainfall.* The normal rainfall of the Amritsar district is 22.24″ and that of the Ferozepur 17.53″. The proportion of irrigated land to total sown acreage is 86.9 per cent in Amritsar, and 57.3 per cent in Ferozepur, according to sample survey estimates.

(iii) *The average size of holdings.* As against an all-India average of 7.5 acre, the average size of holding in the Punjab is 11.8 acres; farms investigated here are all above 8 acres of land; farms are owner-operated except in the case of one farm where some land is leased in.

(iv) *Labour supply.* (i) The most important source of labour in the region is the producer's family. (ii) 10 per cent of the total rural families in the Punjab belongs to the category of agricultural

[6] The data were collected by E. F. Daniel, an agricultural economist from the U.S.A., who spent two years in India (1956-58), assisting the Directorate of Economics and Statistics, Ministry of Food and Agriculture in the analysis of data relating to farm management studies and farm planning projects.

[7] Our programming models were solved at the London Computing Centre.

labour families and the sale of labour to other farms is the main source of income; partly such families are "attached" permanently to bigger farms.

(v) *Resource constraints used in the programming models are*:

(*a*) Land. (i) The size of holding—The statistics of individual holdings are available in acres from the records of Cadastral surveys. A small deduction is made from an individual's total holding to allow for family homestead, boundaries, etc.
(ii) Types of soil—Since the land is of the alluvial type, only one type of land is assumed.
(iii) Seasonal patterns—In the area investigated, land is suitable both for Kharif and Rabi (Monsoon and Winter) cultivation with major crops, if irrigation water and manures are available in sufficient degree. Hence, two land constraints—Kharif and Rabi—have been used, even if the soil type is the same. Assuming that the use of fertiliser and the availability of sufficient water creates possibilities of shortening the period of time involved in maturing crops, we have ignored the possibility of Kharif and Rabi seasons overlapping each other.
(*b*) *Labour* :
(i) Type—The types that have been used here are male, female (family), male (permanently hired).
(ii) Amounts—The data used are of the following types: (I) Data secured from farmers on the number of adult male workers —family and hired — and their working time. (II) Data secured from farmers on the working time of female labourers. (III) The actual working time as existing on the farms is expressed in terms of M.W.D. The total labour-time is expressed as Man Work Days of 8 hours a day and 25 days a month.
(iii) Intensity of work—Since the differences in the intensity of work are not readily reducible to any common denominator, we have abstracted from the problem.
(iv) The choice of slack and busy months—The existing farm plans on the farms concerned are not constrained by the shortage of labour in any season except one farm in one month only. We have, however, used 'labour constraints' in our models since

we are allowing the possibility of using chemical fertilsier which raises labour requirements in different operations.

The factors determining our choice of busy and slack months are (*i*) types of seasons, (*ii*) number of seasons involved, (*iii*) nature of crops technically feasible for each season.

In the present models, the months of October and November have been chosen because the Kharif harvesting and Rabi sowing tend to coincide, land being suitable for both the crops during one agricultural year. The months of June, July and September have been chosen, because Kharif crops (the monsoon crops) require more labour (both for sowing and weeding) than Rabi crops and, therefore, are more relevant for the purposes of formulating labour constraints. The month of August was given up after a few trials when it was found that optimum programmes were not involving any significant reduction in the existing surplus labour in this month.

(*c*) 'Capital or credit' as a constraint—On the farms concerned, very little capital is used for the purchase of variable inputs such as cowdung and seeds. In the existing system with the given technique, farm operations need little short-term credit since the proportion of the purchased inputs to total inputs is extremely low, labour and livestock being the main factors of production.

Hence we have ignored 'credit'[8] as a constraint in our models. However, the credit needs of the farmers might go up significantly

[8]"Capital" in financial terms is sometimes regarded as a "constraint." For instance, Desai has used "capital" as a constraint in his linear programming models on individual farms. "Capital" in his examples means the greater of the two figures A-1 and A-2.

A-1 includes, total hire charges for human and bullock labour, value of owned and purchased manures, seeds, value of fodder, depreciation of bullocks, etc.

A-2 includes, wage cost, seed cost, land revenue, land rent, purchase cost of animals, implements, value of payments to artisans in kind, etc.

"Capital," in the above sense, according to Desai, is a limiting factor on the adoption of optimum cropping patterns. From the standpoint of the optimum activity planning in the short period, however, 'credit' in the above sense has little meaning.

See D. K. Desai, *Increasing Income and Production in Indian Farming*, The Indian Society of Agricultural Economics, Bombay, 1963.

if they use new capital inputs such as chemical fertiliser. We intend to estimate such new credit needs from the optimum programmes based on the use of new variable inputs and draw the relevant policy implications, i.e. relevant for a situation when technical changes are taking place in agriculture.

As regards credit for known (fixed) investments, this has been incorporated in the land constraint.

(*d*) Cattle constraint—The reasons for omitting bullock power as a constraint in the present cases are two: (*i*) it is unlikely to act as a constraint as the farmers concerned have now sufficient bullock power (e.g. some of the farms have 4 or more work animals), (*ii*) even if it acts as a constraint on future programmes, we could not measure it, given the existing data. The sources of data used in our models are given below.

(1) Labour coefficients of crops produced: Regarding indigenous processes currently in use, the labour requirements per acre for crops (existing processes) were collected and summarised by the Farm Management Research Centre in the Punjab state, Ferozepur and Amritsar districts.

So far as the new processes of the existing crop alternatives are concerned, the physical input-output data have been estimated by Daniel on the basis of the results obtained from research experiments conducted by institutions such as the Indian Council of Agricultural Research and information from cultivators in the area based on their own personal observation and experience (e.g. one of the groups of farmers concerned has already started using fertiliser in a small amount).

In the case of the processes based on the use of chemical fertiliser, increased labour inputs required per unit of activity are specifically taken into account.

Although the use of input-output coefficients as described above involves many problems, such use can be defended on the ground that data collected by a research organisation for a period of time from individual farmers concerned are, in fact, likely to be more reliable than those collected by an individual researcher spending a short time in the field and often without much practical knowledge of the region concerned. The use of "experimental" data for new processes in our case is due to the fact that the actual farm data will be available only after producers start adopting innovations, e.g. chemical fertilisers. In the present context, McFarquhar's[9] discussion on the relative merits of the alternative methods of collecting input-output data is of particular significance. (2) Possible crop alternatives: The existing cropping pattern of individual producers is taken as the model with a few modifications: (i) a few "unimportant" crops are omitted—importance being defined in terms of market advantages and the share in the existing cropping pattern ; (ii) livestock activities are omitted: the reason being the absence of physical input-output data;[10] (iii) in order to take into account the non-linear relationships between fertiliser and crops, the total production possibility curve in the case of new processes has been broken up at various levels and different dozes of fertilisers have been treated as independent activities, e.g. a/s 100 wheat, a/s 150 wheat, etc. (iii) Prices of products: Output per unit of activity is evaluated at market prices.[11] The market prices are determined by stuyding the actual prices in the area for a period of 3 years and considering the trends of price movements.

Criticisms may arise here: First, since agricultural producers in India retain a large (though varying) proportion of their total output for self-consumption, market prices cannot be taken as given. Secondly, if all the individual producers start operating at optimum levels, the market prices for their crops will decline, disturbing optimum programmes based on a given set of price-ratios.

As regards the first objection, it might be pointed out that we are here interested in the optimum activity pattern for individual farmers and we have evaluated the net return from all

[9]A. M. M. McFarquhar, "Practical Use of Linear Programming in Farm Planning," *The Farm Economist*, Vol. IX, No. 10, 1961.

[10]We have excluded, from our models, cattle constraint, livestock activities and any specific constraint arising from the need to produce a specific amount of fodder in farm programmes, the reason being the non-availability of data.

[11]The value of by-products such as cotton sticks and wheat *bhusa* has been taken into account; in the region concerned, these products find a ready market.

activities at their market prices. However, it must be acknowledged that predictions on the basis of farm programmes as built here are subject to limitations arising from situations where it may not be economic, from the individual's standpoint, to apply such capital inputs as fertilisers to subsistence crops. Whether such limitations exist and how far they affect the predictions made can only be found out through discussions based on the alternative farm programmes with the individuals concerned.

As regards the second objection, we cannot say *a priori* whether or how prices will go up or down in the specific context of India where there is high income elasticity of demand for agricultural food products today. In any case, we are concerned here with the scope for individual farmers increasing their intensity of operations in the short period. In the short period, the problem, of seasonal price movements, disturbing optimum farm programmes, could be avoided through public policies (i.e. price stabilisation policies).

(*e*) Variable cost items—The relevant costs for the calculation of net revenue that one unit of a particular production process yields to the farm are variable costs, i.e. costs which change with changes in the scale of one activity or process, or when one activity or process is replaced by another.

Items included here in variable costs are: (*i*) Fertiliser cost presents in the case of new processes. The subsidised prices of fertilisers have been used. (*ii*) Seed cost presents in all the cases of cotton, new processes based on chemical fertilisers requiring higher seed rate. The actual prices have been used here.

We have excluded the following items here : (i) Farmyard manure which is produced at the farm, and does not materially affect the choice of any particular activity, and (ii) Irrigation rate which depends in the main on the total size of land to be irrigated and does not affect the choice of any particular activity. (In the case of the individual farms concerned here, irrigation charges act as a "fixed" cost item from the linear programming standpoint).

(iii) Cost of non-family labour. In all the cases investigated here, non-family labour comes from permanent farm employees; hence, the cost of such labour is a fixed item from our standpoint.

(*f*) The objective function and the possible constraints on it.—The models assume the maximisation of family income as the objective of the agricultural producers concerned; the validity of the assumption has already been discussed. (cf. Chapter I).

The possible constraints on the objective function could be:

(*a*) Family's food supply: The subsistence needs (i.e. the food requirement of the family). Formally, such needs can be incorporated in linear programming models by following one of the varied standardised forms.

Here such restrictions are ignored, because, in all cases optimum activities include food grains (specially, wheat) as a major crop.

(*b*) Technically imposed: Considerations of maintaining soil fertility over time often induce a farmer in India to attach greater importance to certain cultural practices (e.g. the mixed culture of a major crop with a legume) than is justified in terms of monetary net return in the short period. Such considerations also inhibit him to resort to double cropping otherwise permitted, unless he is prepared to use extra sources of nitrogen, e.g. night-soil against which he may have personal prejudices.

The use of chemical fertilisers, however, makes it feasible for a producer to escape such restrictions. Since our present study includes fertilised processes for all crops, the above limitations have been ignored.

(*c*) Nature-imposed : Variations in yield arising from the variable nature of the monsoon rainfall affect the choice of crops or activities by producers in Indian agriculture. We abstract from this possible constraint and assume that the individual decision-makers seek to maximise only the expected income (ref. to the assumptions set out in the beginning of the chapter).

(*d*) Institutionally imposed: In the presence of such restrictions as a part of the land not to be 'double-cropped' or 'rent to be paid in the form of a particular crop,' we have to put specific constraints on the objective function; no such limitations arise, however, in our case. Of the five farms taken into account, four are owner-operated. In the case of one farm only, there is land rented on both share and cash basis along with owned land (2.5 acres owned land + 5.5 acres rented in on cash basis + 1.90 acres rented in on share basis). The land reform acts in the region concerned have already removed some of the old abuses of the tenancy system.

Linear programming terminology as used in the present discussion is as follows :

(I) *Objective function:* The objective taken to be maximised in the models, namely, family income in the present cases.

(II) *Optimum programme or activities:* A combination of activities which maximises the objective function given the resource constraints.

(III) *Effective constraints:* Resources which are exhausted in the optimum programme and have positive marginal value productivity.

(IV) *Ineffective Constraints:* The resources not limiting optimum programme and having zero marginal value productivity.

(V) *Marginal value productivity* : The imputed prices corresponding to the marginal productivities may be taken to imply that the imputed value or price of *ith* factor indicates the amount by which net revenue can be increased by adding to the supply of the resource concerned by one unit.

(VI) *Income to fixed factors:* Gross income-variable cost of producing the particular activities included in the optimum programme.

VI: 2 RESULTS[12]

First, in all the six cases studied, income to fixed factors (see Appendix C) increases annually as activities based on chemical fertilisers

[12] See Appendix (B) for the matrices of the programming models and the description of the farms.

replace the existing non-fertilised ones, income being evaluated in both the cases (in the existing and the optimum programmes) at constant prices.

Farms	*% Rise of Income*
No. I	126.2
No. II	81.64
No. III	100.6
No. IV	66.7
No. V (Model I)	146.4
(Model II)	187.0

The differences in the rate of increase in income for the individual producers concerned arise from such factors as the inclusion of low-valued and low-labour-content crops in optimum activities (ref. Farm No. II and IV) the ratio of double-cropping in the existing programme to that in the optimum ones, etc.

Secondly, the rise in income to fixed factors is associated in all the cases, with a rise in labour time utilised on the farms concerned. Since such increased labour utilisation is associated with reductions in the surplus labour time now existing on the farms, the rise in income is based on a rise in labour productivity per head.

Thirdly, the rise in income to fixed factors is associated with a rise in the utilisation of land available to individual farmers.

In terms of our previous discussion (cf. Chapter I), we might argue here that the variation in the proportion of double-cropping to total land available on the farms studied in the existing system may partly be related to the size of the family. More specifically, farmers with large families might give up their concern for maintaining soil fertility over time and resort to an intensive use of land without using extra fertilisers under the pressure of securing greater income for their families. For instance, Farm No. I, having the highest intensity of cropping in the existing system (93.46%) is the smallest in size in the groups and has the largest family (9 members). Farm II is also a smaller farm compared to the others. Farm No. IV though larger in size (13.17 acres) than Farm No. III (12.50

REDUCTION IN THE EXISTING SURPLUS LABOUR TIME

(in percentage)

Farms	*June Labour*	*July Labour*	*September Labour*	*October Labour*	*November Labour*
No. I	0	35	19	100	(not used as a constraint)
No. II	0	20	0	100	(,,)
No. III	86	36	15	(present shortage is over-come and the avail-able supply is fully used)	0
No. IV	57.7	54.8	29.0	100	
No. V (Model I)	0	64.2	23.6	100	(not used in the model)
(Model II)	0	100	100	100	(not used in the model)

NOTE: The reduction in the surplus labour time in the months of July and September represents the increase in the labour requirements for operations such as weeding; in the month of June, for Kharif sowing; months of October and November for Kharif harvesting and Rabi sowing.

Farms	*% of Double Cropping in Optimum Programme*	*% of Double Cropping in the Existing Programme*
No. I	100	93.46
No. II	84	74.47
No. III	100	30.0
No. IV	100	93.77
No. V (Model I)	98	0
(Model II)	49.3	0

acres) has higher intensity of cropping (93.77%) than the other (30%); the family in the case of the former is larger than in that of the latter. Indeed, both the Farms No. I and IV, having the highest intensity of cropping in the existing system, have also the largest families (9 members), while the average for the group is 7.5 members.

The intensity of cropping in the existing system may also be related to the income-leisure substitution rate at the current income level which might be taken to be related to both the size of farm and the size of family, assuming that no change is taking place in the want pattern of the families concerned. Farm No. V is the only farm which does not have any double-cropping in the existing system and this is also the largest farm (18.75 acres of owned land) in the group studied. Again, this is the only farm in the group where no female member of the family works on land[13] and 10.25 acres of land are kept unirrigated (all the other farms irrigate their land fully).

Lastly, the rise in income to fixed factors is associated with a rise in working capital requirement as represented by fertiliser cost.

The above rise in the requirement of working capital[14] on the farms concerned represents a large and sudden increase in the credit

FERTILISER COST

(*in rupees*)

	Optimum Programmes	*Existing Programmes*
No. I	386.25	0
No. II	456.30	0
No. III	591.40	0
No. IV	530.00	0
No. V	720.75	20

NOTE : We have ignored here the rise in the requirement of working capital due to the increased seed costs.

[13] In the region concerned it is customary to have female members of a family working on land. See The Royal Commission on Indian Agriculture, 1928, *Evidence Volume VIII*.

[14] A comparison with the actual experiences in Japan where agricultural technique is oriented towards the intensive use of land based on chemical fertilisers and labour input is relevant here.

needs of the farmers. Not only the amount involved is large, relative to the existing use of credit, but also its nature is different; it is short-term capital need. The existing credit, structure in India is geared so far to satisfy the credit needs for long-term investment or short-term needs for consumption purposes. Private money-lenders supply such loans at very high rates and often secure control over the future disposal of surplus crops by the debtor. Such a credit system is unlikely to be able to meet productive credit needs on a large scale at a known and fixed rate of interest and without involving extra commitments on the part of the debtor. We must acknowledge that it is one thing to raise loans at exorbitant rates from money-lenders to meet the family's basic requirements, or to overcome a contingency arising from crop failures, or the death of cattle[15] or even to meet such sentimental needs as performing the last religious rites for parents; it is altogether a different matter to take loans at such terms to experiment with a new technological innovation, however large may be the promises held out. (see, later discussion on policy implications).

The above summarized results of our programming models give rise to certain implications which we discuss in the next section.

On a Japanese farm in the Kinki District (comparable to some rice districts in W. Bengal), for example, the operating expenses are 8 times as high as on the Indian farm, the cost of fertiliser used being 16 times as high. The economic consequences are : (i) the gross output per acre in Japan is valued at $448.30, and in India at $45.80; (ii) the labour utilisation on the Japanese farm is 3.8 times as large as on the Indian farm, while the stock of labour on the former is 50 per cent larger than that on the Indian farm.

See, J. W. Mellor, "Increasing Agricultural Production in Early Stages of Economic Development : Relationships, Problems and Prospects," *Indian Journal of Agricultural Economics*, April-June, 1962, p. 42.

[15] *Papers and Proceedings of the Deccan Agriculturists Act of 1879*, Government of India, 1882, esp. Minute by the Governor of Bombay, 12*th* November 1877, and T.C. Hope's speech introducing the Deccan Agriculturists' Relief Bill at the Council of the Governor-General meeting on 17 July 1879; The Famine Commission, *Report*, 1901. The Royal Commission on Indian Agriculture, *Final Report*, 1928; *Revenue and Accounts, Government of India, 1902*, Sir M. Darling, *The Punjab Peasant*, 4th edition, specially p. 222. According to the Rural Credit Survey, 1954, 47% of loans taken by small farmers is devoted to meet consumption needs of families on an all-India basis, while there are wide regional variations, see *Report*, Vol. I, Part I, p. 265 and Vol. III, p. 587.

As to the question whether they could be implemented on the farms concerned, no *a priori* answer is possible without actual discussions with the farmers. Two points might be made here :

First, our simplified normative farm programmes might need adjustments through discussions with the individuals concerned. The precise nature and directions are likely to vary from farm to farm. *A priori*, it might be said that the implementation of such programmes will depend on the producers' own risk-aversion and their preparedness to give a trial to the programmes.

Secondly, the farm plans might be sought to be implemented only if the assumptions made in the beginning of this section hold good. For instance, information about chemical fertiliser, the complementary inputs, e.g. water, credit are available; the market prices do not fluctuate creating new risk elements in the farm-planning environment.

We might conclude, given the above factors, that these farm plans might be implemented with the given land and labour resources available on the farms concerned. This assertion is based on the following arguments:

First, the farms we have studied are situated in a region where earlier attempts at spreading innovations (e.g. American cotton, improved seeds, canal irrigation) were fairly successful. All such innovations raising yield per acre of land were incorporated into farm planning.

Secondly, the optimum cropping pattern based on chemical fertiliser might be regarded as not spectacularly different from the existing ones. For instance, multicropping, which is a universal feature of the existing farm plans in the region, is an important feature of the new activity pattern as well. Similarly, crops which play relatively important roles in the existing system, namely, cotton and wheat, enjoy similar importance in the new pattern[16] (see Table 15).

[16] It might be acknowledged here that minor adjustments in the cropping pattern may be necessary to satisfy specific constraints, e.g. the supply of a minimum amount of fodder, after discussions with individuals concerned.

TABLE 15

CROPPING PATTERN : OPTIMUM PROGRAMMES ON FIVE INDIVIDUAL FARMS IN THE PUNJAB

(*in acres*)

	Farm No. 1	*Farm No. 2*	*Farm No. 3*	*Farm No. 4*	*Farm No. 5 Model I*	*Model II*
1. Cotton A,K, a/s 100	8.50	1.49	6.72	3.25		
2. Cotton A,K, a/s 150					8.42	8.75
3. Fodder K, a/s 100		5.06				
4. Maize (K, mix 100)				4.29		
5. Vegetable (K, mix 100)			5.78			
6. Bajra* (K,FYM)					0.08	
7. Fodder** (K,R, mix 100)				2.32		
8. Vegetable (K,R,***mix 149)				3.31		
9. Wheat-Gram (R, mix 100)	8.38				8.50	17.50
10. Wheat (R, mix 100)	.22		12.50	7.54	10.25	
11. Wheat (R, mix 130)		9.60				
12. Fodder (R, mix 100)						0.80
13. Paddy (K, a/s 100)		1.13				

*On unirrigated land (mix 20).
**An indigenous process.
***Covering land for both the seasons K and R.

Thirdly, the region concerned has rich soil found to be specially responsive to the use of chemical fertiliser; much of the land is already irrigated while new irrigation projects and a new fertiliser factory (at Nangal) are in the process of completion. An assured supply of fertiliser and water, under given circumstances might not be as impracticable as is often assumed.

Fourthly, the most important argument in favour of the view that the above farm plans might be implemented in practice is that it could be done with the already available supply of labour-time on the farms concerned. This implies that the introduction of chemical fertilisers will enable these family farmers to raise income to the existing fixed factors, i.e. land and labour. The production possibility envelopes will be pushed outward and the result will be a rise in the productivity of both the major resources now available, namely, land and labour.

So far as the last factor is concerned, it might be regarded as the fundamental characteristic of the majority of family farms in India. The majority of the small family farms appears to have more labour-time available than could be utilised even in periods of peak demand for labour in agricultural operations. So far as this characteristic is of reasonable validity, the farm plans develpoped in the present case have also important implications for other regions where water and fertiliser could be supplied.

VI: 3 Implication

The implications drawn from the above study are as follows: First, any change in the existing farming system increases the capital need of the family farmers concerned. No doubt, part of the need could be met by the farmers concerned out of their family savings (cf. Chapter II). But in practice, such possibilities will depend on the actual income level of the farmer concerned and on his willingness to part with liquid cash resources. In India, agricultural producers appear to have a strong preference for liquid resources such as cash or gold,[17] given the importance of risks arising from the

[17] Gold and silver have always been important sources of liquid assets in India (e.g. the free minting of these metals continued as late as the end of the 19th century). A recent attempt at estimating the total stock of precious

fluctuations in rainfall and the repeated occurrence of such natural hazards as 'floods' or 'droughts.' Whether their concern for raising family income will induce them to part with their liquid assets at a point of time cannot be answered *a priori*. It has to be enquired into empirically. Moreover, the producers concerned may find it more attractive to invest family savings in 'known' inputs such as land and cattle rather than in chemical fertiliser which is still a new input. Again, what an individual will actually do, faced with a choice, cannot be said *a priori*.

From our standpoint we can say one thing, *a priori*, namely, that the credit policy of the public authorities should be geared to meet the requirements of working capital which arise as a logical consequence of any efforts to spread innovations—embodied in variable capital inputs in the existing system. Such problems as, what institutions should be used to supply credit; what should be the security asked for such loans, what should be the rate of interest rise in this context. Although the present study cannot offer any definite answers to any of the above questions because they require specialised independent enquiries into the problems involved, some suggestions might be noted here. Since the gestation period involved in loans to finance the purchase of chemical fertilisers is short (i.e. one season of crop production), such loans might be offered to the individuals willing to introduce innovations without asking for any security. As a protection against the ineffective use of such innovations, the credit institutions may associate cash outlay with instructions on how to use the new inputs. A credit-cum-managerial institution can help producers to use new inputs in the best possible way while minimising the risk of loss involved in loans without security. It must be emphasised that producers' initiative in experimenting with new inputs may often be vitiated

---

metals is made by V. Prakash. See *Papers on National Income and Allied Topics*, ed. V. K. R. V. Rao, S. R. Sen *et al.*, Vol. I, 1960.

That the agriculturists use such resources to tide over economic distress caused by Nature's caprices has been demonstrated over time. For instance, the average monthly value of ornaments sent to Bombay Mint jumped to Rs. 1,900,000 in September 1877 from a current average value of Rs. 6,000; the year 1877 was a year of famine in the province and the adjoining areas. See *The General Report of the Indian Census*, 1891, p. 24.

by their failure to secure as much benefit as expected due to their inefficiency or the lack of knowledge. Further, credit institutions can provide loans not in the form of cash but in the required amount of fertiliser as such. Such loans *in kind* associated with the supply of information as how to use them in the best possible way might go a long way to assure farmers of the reliable supply of new inputs at a known price and without involving any extra commitments. Though chemical fertiliser epitomises innovations in our case, the above argument applies to any other new input.

The second implication following from our study is that when technological innovation is directed to raise yields per acre of land mainly through the use of 'biological' inputs (e.g. chemical fertilisers, hybrid seeds), the demand for labour goes up in the agricultural sector. For instance, in the case of the farms studied here, the existing surplus labour-time is reduced in more than one month, while the existing labour supply is exhausted in one or two months. Thus, the 'labour surplus' on the family farms studied appears to be a phenomenon linked up with the existing technique in agriculture which uses both land and labour inefficiently.

This explains why our programming results differ from those of Desai, for instance. Desai[18] concludes on the basis of his 20 programming models on individual farms in the Bombay state, Ahmednagar District, that as many as 6.87 adults or 16.5 per cent of the total working force "surplus," i.e. not required in farm operations in any month of the year, even if optimum programmes are introduced. Desai's farm programmes are built on the assumption of the existing technology continuing in the future. In our case, family labour will remain under-employed in the slack months, but all the existing workers are required on farms in busy months. In the respect that the demand for labour goes up with improved technique, our results bear out a similar conclusion reached by Daniel on the basis of his study of six individual farms in each of five different states in India.

Some objections may arise against the above argument :
First, the labour-time taken to be available in our models is based

[18] D. K. Desai, *Increasing Income and Production in Indian Farming*, The Indian Society of Agricultural Economics, Bombay, 1963.

TABLE 16

LABOUR INPUT-OUTPUT RELATIONSHIP: STATE AVERAGE, BASED ON SELECTED FARMS

State	*Man Available*	*Work Days Used* *P*	*A*	*Output per Man Work Day* *P*	*A*	*P.C. Increase in Output*
W. Bengal	940	448	667	2.23	3.54	59
Madras	855	380	546	2.19	3.07	40
Uttar Pradesh	918	705	819	2.48	3.87	56
Bombay	894	566	736	1.87	3.44	84
Punjab	1,093	747	913	2.52	3.95	57
Average	940	569	736	2.26	3.57	58

P=Present Plan *A*=Alternative Plan

The additional returns to labour shown here resulted from improved technology under good management and not from improved labour methods or the introduction of improved tools.

SOURCE: E. F. Daniel, *Farm Planning and Management*, Directorate of Economics and Statistics, Ministry of Food and Agriculture, India, 1960, p. 269.

on the present working pattern and underestimates the feasible labour-time available on the farms concerned. In the existing family set-up in India, family members often split up work in order to give each individual member a social recognition for sharing in the total produce of the family. The actual working time at present, given the narrow range of technical possibilities and the absence of alternative occupations, may represent neither the "leisure-work preference" of the producers' families nor their "ability to work"; it may be a simple reflection of an arbitrarily imposed leisure on the individuals due to population pressures and the lack of alternative occupations.

The actual possibilities of such "unused" labour-time (family), however, in the case of the present farm units, are small, because, male adult members are already working more or less full time in these family farms which are larger than the all-India average holdings, more or less irrigated fully (e.g. Farm V) and have double cropping.

In the case of female labour, some time will have to be devoted for household works (at present, female labour is working on a part-time basis on all the farms investigated except the largest one).

The inclusion of child labour, we feel, will be undesirable on social grounds. Lastly, some margin of 'surplus' labour-time on family farms or labour hoarded (i.e. labour time which is not in use every season but can be drawn on, if required) is needed to meet the effects of unforeseen circumstances, e.g. illness, yield fluctuations. For instance, the employment of human labour per acre of irrigated land under wheat-gram mixture may increase by 30 per cent for preparatory tillage, and by 77 per cent for reaping and threshing in a year of good rainfall in the Amritsar District where farms studied by us are situated. The greatest source of such of the 'surplus' labour-time on family farms is the flexibility of working hours for family members and farm labourers annually attached to farms. At critical times, family members are often found not to take time off from work on any particular day and the traditional religious festivals in villages are concentrated in slack seasons.

Secondly, our sample is very limited, i.e. has only six individual farms; hence, no indication as to the rise in demand for labour in agriculture, on an all-India basis, could be deduced from our study.

Despite this obvious limitation of the present study based on a limited sample, some light is surely thrown on the problem of "labour surplus" in general. The reasons are the following:

(*a*) large possibilities of double-cropping exist in other parts of India as well; a part of which will be practicable if the irrigation targets of the Third Five Year Plan are fulfilled.

(*b*) In the case of rice tracts (i.e. rice followed by rice), more than one crop add more to labour requirements than in the wheat tracts (to which our present farms belong) where lower labour content Rabi crops (wheat) follow higher labour-content Kharif crops (cotton).

(*c*) In our models, we have taken into account only chemical fertiliser as an innovation; the introduction of other proposed changes, e.g. improved seeds, the biological and the cultural controls of insects and diseases, or the better preparation of the farmyard manure, will all add to labour requirements. No doubt, much of the increase in labour requirement due to the above innovations will take months, having surplus labour-time in our models, namely, July, August, September, but the demand for labour at critical moments, i.e. harvesting, will also arise due to larger crops.

(*d*) In our models we have ignored the livestock industry; the better utilisation of the existing stock of this resource will increase further labour requirement on family farms.[19]

In the present context, a distinction between 'labour-time' and 'labourer' must be made. An increase in the demand for labour, as noted in our cases, does not necessarily imply an increase in the demand for workers. Much of the increased demand for labour is in the months when family members and other permanent workers have very little to do in the existing system. Demand for labour goes up also in those months where the existing workers are occupied (e.g. times of harvesting); nevertheless, the rise in demand for extra labour-time, for a period of 15 days in a year, for instance,[20] is unlikely to induce a decision-maker to raise the number of permanent workers who have to be paid wages throughout the year. In the case of family members, it implies an extra mouth to feed throughout the year, while the contribution made by the member concerned is only for a short period. Such need for extra labour-time is met in other agricultural systems either by seasonal employment of workers or by the existing workers, working overtime or by mechanical improvements. This explains why brisk work at harvest times or even acute shortage of labour at times of peak demand (e.g. harvesting) may coincide with a situation where

[19] In India, there are possibilities of developing the livestock industry. Two factors must be noted in this connection. First, the average income elasticity of demand for dairy products is high, e.g. the estimates of income elasticity of demand for milk products are 1.83 in rural and 1.67 in urban areas. (See *4th, 5th, 7th, 9th* and *10th Rounds of the National Sample Survey.*) Secondly, the traditional pattern of farming was geared to joint crop and dairy activities. *Problems of Food and Agricultural Expansion in the Far East*, FAO, 1955.

[20] To suggest to an individual producer that he should engage an extra labourer permanently to meet this demand is like suggesting to a big department store to determine its number of workers all the year round, by the peak demand for 'labour' at the Christmas-week shopping.

agricultural households might still like some of its members to go out and find alternative productive employment. In India, the presence of agricultural labour families who do not find productive employment in the busy months, and of the small farmers having less land which could yield bare subsistence, the increase in labour requirement due to improvement in techniques might go without any increase in the number of workers employed in agriculture. As noted earlier, in Japan, a 2 per cent annual increase in the output, was achieved by a constant labour force through individually taken decisions by family farmers.

A third implication drawn from our present investigation is that, given the land and labour supply on the existing farms, the higher marginal value productivity of land[21] (e.g. MVP of Rabi land moving between Rs. 92.52 and Rs. 159, of Kharif between Rs. 13.64 and Rs. 115.66, of October labour between Rs. 0.39 and Rs. 15.07) might be taken to throw some light on the direction of new investments in agriculture in the future (see Table 17). It is very likely that the new investments in agriculture will take the form of land-saving devices, at least in the case of farmers having less than 20 acres[22] of land (i.e. 96 per cent of agricultural producers). This conclusion appears to be consistent with the general consensus of opinion regarding the nature of future investments in Indian agriculture.

The above argument is, however, subject to three qualifications: (i) On the large farms, labour saving devices may be emphsised more than the land-saving ones, depending on factors such as the marginal value productivity of the inputs concerned, wage-rate of casual labour, etc. (ii) On small farms, the solution to seasonal shortages of labour might often be the introduction of improved tools which will not only release family labour at critical times but will also enable them to work more efficiently (cf. Chapter

[21] Cf. R.J. Freund, *Introduction of Risk into a Linear Programming Model*, Unpublished Thesis, Microfilm, British Library of Political and Economic Science, London School of Economics; also S. Chakravarty, *The Use of Shadow Prices in Programme Evaluation*, Economic Development Programme, India Project, C/60-15.

[22] The largest farm studied here has 18.75 acres of land; this aspect of the problem has been investigated parametrically in the next chapter.

V). (iii) The saving of land or labour in agriculture is possible also through 'disembodied' investments, i.e. investments which do not involve any fixed capital expenditure.

TABLE 17

SHADOW PRICES OF THE SCARCE RESOURCES BASED ON OPTIMUM FARM PROGRAMMES

(in rupees)

	Farm No. 1	*Farm No. 2*	*Farm No. 3*	*Farm No. 4*	*Farm Model I*	*No. 5* Model II*
1. Irrigated Kharif Land	115.66	0	109.0	13.64	64.50	0
2. Irrigated Rabi Land	157.16	104.54	158.95	92.52	150.53	101.50
3. Labour July	0	14.79	0	0	0	1.35
4. Labour September	0	0	0	18.56	0	17.03
5. Labour October	6.83	0.39	8.5	6.71	15.07	10.22
6. Labour November	N	N	0	.95	N	N
7. Labour June		5.62				

N=not used as a constraint in the particular case.
*Land unirrigated: Rs. 73.91.

For instance, a change in the irrational size, shape and layout of farms, as determined by the compulsions of property divisions[23]

[23]According to the findings of the National Sample Survey, July 1954-March 1955, 8th Round, only 8% to 9% of the total cultivated land is jointly operated in India, the rest being individually cultivated. In many countries, legal and social steps have been taken to prevent the division of agricultural land among many heirs. For instance, the Austrian Act of Inheritance grants farm land to one heir; special credits are made available in Japan to enable a single heir to take over the entire property; in Egypt, a single heir is to be determined by drawing lots among claimants.

among many heirs may save not only labour but also land.[24] Similarly, the modern technique of the time and motion study, simplifying body motions and eliminating unnecessary walking, may reduce the wastage of labour-time on farm units while involving no fixed investments.

The fourth implication drawn from this study might well be taken as an answer to a question raised at the end of Chapter IV, namely, whether technical changes provide a way of raising family income for individual decision-makers in Indian agriculture. Our enquiry into the cropping pattern of two individual producers in the Punjab showed that if they try to move up the EV curve to raise their family income in the existing system, they raise income variability by many times more than the rise in the expected income (cf. Chapter IV). Such movements were seen to involve greater specialisation in crops with high income variability. The present discussion shows that if innovations are selected with full recognition of the complementarity among inputs, family farmers might have an opportunity of raising their family income without increasing risks involved in the farming environment to any significant level. This opportunity, however, depends on the fulfilment of a large number of conditions, a few of which are particularly emphasised here :

(*a*) The availability of new inputs such as chemical fertiliser

[24]The idea often put forward that the field boundaries due to subdivision and fragmentation do not constitute any 'real' problem, since many agricultural products need such bounding for holding water is hard to accept. The size, shape and the layout of plots bounded on technical grounds need not conform to size, shape and layout, resulted under the operation of laws of inheritance giving equal share to all heirs and operating in a socio-economic environment where all heirs stay on land and each seeks a slice of each different type of land (fertility and location) owned by the father. In any case, rice is the only major crop requiring small plots bounded on four sides.

According to the Punjab Administration Report for the year ending 30th September 1955, the consolidation of the fragmented holdings resulted in an increase of 59,687 acres of cultivable land (innumerable boundary bunds were removed). The consolidation of holdings also reduces labour demand on farms as unnecessary travels are reduced and the unit of operation increases.

See *Consolidation of Holdings—Methods and Problems*, Planning Commission, Government of India, Delhi, 1958, pp. 72-9.

and hybrid seeds must be associated with the assured supply of sufficient water.

(*b*) The prices of new inputs and the prices of products must not fluctuate too much in the short period.

(*c*) The public authorities must be prepared to satisfy credit needs of farmers arising from their willingness to use new inputs at known and fixed rates without creating administrative delays, which very often discourage producers from going to organised institutions for loans.

(*d*) The public authorities should refrain from creating risk situations through inconsistent and contradictory legal and administrative measures interfering with the decision-making processes on farming units. Moreover, the public authorities must be willing and able to remove constraints arising from the existing organisational structure, e.g. land tenure system, marketing arrangements, etc. such requirements might vary from region to region in importance.

It must be acknowledged in conclusion that the effectiveness of new capital inputs in raising output and income, in the short period, is determined largely by the human factor involved in family farms. The variations in yield per acre of land in the existing system, given the same technique and the same soil-climatic nexus, is often due to the human factor involved. However, even this essentially non-measurable variable could be influenced through education, training, and perhaps by creating an environment where an individual has the opportunity to improve his own and his family's lot if he so desires.

Another final implication follows from our study. Statistical data, so far collected, are insufficient to build up alternative farm plans for individual holdings. New inputs are emphasised; but their implications for the labour requirement or for capital requirement are often difficult to measure today. In the short period, policy implications could be better formulated if alternative farm plans, incorporating innovations, are made and then discussed with individuals concerned to assess the possibilities of their implementation realistically. Through such assessments we could locate factors inhibiting individuals even when they are willing to give trial to new plans. Such enquiries need field investigations in different regions in the country. Any realistic plan for raising agricultural output in the short period, must take into account actual constraints which might discourage producers from changing their existing systems. This need cannot be over-emphasised in the context of India where farm decisions are taken individually on family farms.

CHAPTER VII

EFFECTS OF LAND-LABOUR RATIO IN OPTIMUM FARM PLANNING

IN THE previous chapter, we have discussed how a group of individual farmers might raise their output and income by utilising their fixed resources, namely, land and family labour in a more efficient way than in the existing system.

The use of new capital inputs, epitomised in chemical fertilisers, makes possible such improvements in the use of resources. We noted also that, given the relative marginal value productivity of fixed resources, the farmers concerned might find it economical to invest in innovations which save land more than labour.

In the present chapter we examine, how the utilisation of fixed resources (i.e. land and labour), the cropping pattern, the rate of increase in income to fixed factors change with a change in the supply of that fixed factor which appears to be the factor with the highest marginal value productivity, namely, land. From such an investigation we would be able to derive also that scale of operation where the land available could be used fully with the given supply of labour. We noted earlier (see Introduction) that in the Indian context, the full utilisation of land available might be an objective of macro-economic planning. Here we intend to assess the implications of the above objective in terms of individual farm planning.

In the first section of the present chapter, we build up new farm programmes parametrically and discuss the changes in the resources utilisation pattern, the choice of the crop alternatives, and the possibilities of a rise in income to fixed factors as embodied in the optimum programmes. In the second section, we discuss some of the conclusions which might be reached on the basis of the present study.

VII: 1 MODELS

The tool of investigation used in this section is linear programming with variable resource supply where we have allowed the land available to vary, within a range, while keeping the given family-labour size.

In such investigations that scale of operation, at which the land is fully utilised with the given labour supply and beyond which land remains unutilised, can be located, since each optimum programme with varying land resources gives us the amount of this resource as utilised at each level. The ranges below and above that scale are investigated to find out, (i) what happens to family income, the cropping pattern, and the resource utilisation when land is the "limiting factor"; (ii) what happens to the above variables when land is not the "limiting factor."

Our programmes with variable land supply have been solved on the Mercury computer at the London University Computing Centre.

The individual farms investigated for the above purpose are: (*a*) Farm No. 5, and (*b*) Farm No. 3 (see Chapter VI and Appendix B).

The reasons for the choice of Farm No. 5 are : (i) this is the largest farm in the group studied; (ii) our ordinary linear programming models (see Chapter VI) have already shown that the farmer concerned cannot utilise his land fully with technological innovations being introduced, since labour at critical times is a limiting factor.

The reasons for the choice of farm No. 3 are that : (i) its total labour supply represents one of the largest labour size in the group of farms investigated earlier; (ii) its cropping pattern includes 'vegetables' which offer a lower labour-content crop alternative to 'cotton' in the Kharif season.

A comparison of the farm programmes on the above two farms might indicate how the full utilisation of land on invididual farms is related not only to the available labour supply but also to the nature of the crop alternatives. The use of land on an individual farm unit might be increased, with the given labour supply, by creating new crop alternatives which require less labour than the existing ones.

The resource availability of the two farms concerned here is as follows:

(*a*) *Farm No.* 5. (See Appendix B):

(*i*) Labour Supply: 105 M.W.D. every month.

(*ii*) Land Supply : The actual land-holding is 18.30 acres, for the parametric investigation the availability of land (irrigated) has been allowed to vary from 10.25 acres.

(*b*) *Farm No.* 3. (See Appendix B):

(*i*) Labour Supply : 95 M.W.D. per month.

(*ii*) Land Supply: The actual land-holding is 12.50 acres, for the parametric investigation land (irrigated) has been varied from 8 to 25 acres.

Regarding other data, e. g. input coefficients, costs and returns of each activity, we have used the same data as used in earlier programming models.

The results of the linear programming (parametric) models are summarized below.

(*a*) *Farm No.* 5 (See Table 18):

(*I*) *Land utilisation* (*see Graph* 3):

(i) With the given size of labour supply, the optimum scale of farming is 12 acres of irrigated land.

(ii) On land bigger than 12 acres, the farmer keeps more and more Kharif land unutilised.

(iii) In the range of 12-19 acres, the proportion of the non-utilised Kharif land increases with a rise in its supply.

(iv) Beyond 19 acres, any further increase in the supply of Kharif land simply adds to the amount of non-utilised Kharif land.

(v) Rabi land remains utilised till its size reaches 25 acres (i.e. the limit of the size variations allowed in our models).

(*II*) *Labour utilisation.* (*see Graph* 4):

(i) On 12-acre land, July and October labour emerge as the limiting factor.

(ii) In the range of 13-25 acres, July, September and October labour remains the limiting factor.

TABLE 18

RESULTS OF PARAMETRIC PROGRAMMES: FARM No. 5

Land size (acres) Double cropping	*Cropping Pattern*		*Bajra (R) (Indigenous)*	*Wheat-Gram (R) (mix-100)*	*Fodder (mix 100)*	*Surplus Land (Kharif) (acres)*	*Total Income (Rs.)*
	Cotton (A.K.) (a/s 100)	*Cotton (A.K.) (a/s 150)*					
	(acres)	(acres)	(acres)	(acres)			
12	8.33	1.17	2.49	12.00	0	0	4130.07
13	9.43	0	2.56	13.00	0	1.00	4276.20
14	7.37	1.42	2.10	14.00	0	2.60	4393.80
15	5.53	1.42	1.50	15.00	0	4.35	4503.90
16	3.37	5.61	0.90	16.00	0	6.12	4614.04
17	1.12	7.70	0.30	17.00	0	7.87	4724.18
18.3 (actual)	0	8.75	0	17.50	0.80	9.55	4860.45
19	0	8.75	0	17.50	1.50	10.25	4931.50
20	0	8.75	0	17.50	2.50	11.25	5033.00
21	0	8.75	0	17.50	3.50	12.25	5134.50
22	0	8.75	0	17.50	4.50	13.25	5236.00
23	0	8.75	0	17.50	5.50	14.25	5337.50
24	0	8.75	0	17.50	6.50	15.25	5439.00
25	0	8.75	0	17.50	7.50	16.25	5540.00

(*III*) *Cropping Pattern* (*see Graph* 5):

(i) On Kharif land, the process for American cotton changes from the one using a/s 100 fertiliser to a/s 150 fertiliser as the land size increases from 12 to 13 acres; the latter replaces the former at 18.3 acres size and beyond.

(ii) The indigenous process for bajra, a Kharif crop, remains in the optimum programme till the size of holding reaches 17 acres (this is the only case of an indigeneous process remaining in the optimum programme when fertilised processes are allowed in the model).

(iii) Beyond 18.3 acres, *the whole* of the utilised Kharif land gets *stabilised* at 8.75 acres, devoted to cotton (A) (mix 150).

(iv) On Rabi land, wheat-gram mixture covers the whole land supply till the total land size reaches 17.50 acres.

(v) Beyond the range of 17.50 acres, any further addition to Rabi land, adds to the acreage under fodder.

In all the above changes in the cropping pattern of farm No. 5, the crucial consideration appears to be to save labour at critical times by resorting to low-valued and low-labour content crops in the place of high-valued, high-labour content crops. For instance, bajra (indigenous process) included in the optimum programme, requires less labour than bajra (fertilised) which is excluded; again, the labour coefficients of bajra as such are lower than those of cotton with which it competes for Kharif land in the crucial months of October and September. Similarly, fodder competing with wheat-gram mixture for Rabi land requires much less labour than the latter, specially at critical times.

(*IV*) *The increase in income to fixed factors:*

(i) The increase in income slackens as the size of land-holding goes up; for instance, income falls from 3.53 per cent between 12-13 acres, and 1.8 per cent between 24-25 acres.

(ii) The rate of growth in income is much higher when the farmer is changing from the existing indigenous processe to new processe, based on fertiliser (see Chapter VI).[1]

[1]This low percentage increase in income to fixed factors arises from two factors : (*a*) the fall in the intensity of cropping as the land size increases, (*b*) low-valued crops replace high-valued ones.

Table 19

RESULTS OF PARAMETRIC PROGRAMMES: FARM No. 3

Land Size (*acres*) (*Double Cropping*)	*Cropping Pattern* *Cotton* (*A.K.*) (*a/s 100*)	*Vegetables* (*K. mix 100*)	*Wheat* (*R. mix 100*)	*Fodder* (*K.R. a/s 100*)	*Surplus Land* (*Rabi*)	*Total Income* (*Rs.*)
	(acres)	(acres)	(acres)	(acres)	(acres)	
8	8.00	0	8.0	0	0	2911.99
9	8.16	0.84	9.0	0	0	3219.05
10	7.75	2.25	10.0	0	0	3486.99
11	7.38	3.66	11.0	0	0	3754.94
12	6.92	5.07	12.0	0	0	4022.90
13	6.50	6.50	13.0	0	0	4290.85
14	6.10	7.90	14.0	0	0	4558.80
15	5.00	10.00	15.0	0	0	4780.00
16	3.75	12.25	16.0	0	0	4991.00
17	2.50	14.50	17.0	0	0	5201.99
18	1.85	16.75	18.0	0	0	5412.99
19	0	19.00	19.0	0	0	5623.99
20	0	19.08	18.33	0.91	0.75	5736.58
21	0	20.11	17.66	0.88	2.45	5745.60
22	0	21.15	17.00	0.85	4.15	5754.65
23	0	22.18	16.33	0.81	5.85	5763.68
24	0	23.22	15.66	0.78	7.55	5772.70
25	0	24.25	15.00	0.75	9.25	5781.75

(*b*) *Farm No.* 3. (*see Table 19*):

(*I*) *Land Utilisation* (*see Graph 6*):

(i) Land is fully utilised (i.e. double-cropped) with the given supply of labour (family labour only) till the size of holding reaches 20 acres.

(ii) On land more than 20 acres, the farmer keeps Rabi land unutilised. The proportion of Rabi land unutilised increases with a rise in supply beyond 20 acres.

(iii) On 25-acre land, the total amount of unutilised Rabi land is 9.25 acres.

(iv) Kharif land remains utilised up to 25 acres (i.e. the limit of our parametric investigations).

(*II*) *Labour Utilisation* (*see Graph 7*):

(i) On 10-14 acres land, October labour is a limiting factor (i.e. the supply is exhausted).

(ii) On 15-19 acres land, June labour is the limiting factor.

(iii) On 20-25 acres land, the available supply of the June and November labour is exhausted in optimum plans.

(*III*) *Cropping Pattern* (*see Graph 8*):

(i) On land more than 8 acres, any addition to Kharif land adds to the acreage under vegetables. On 9-acre land for example, vegetables cover 0.84 acres and cotton 8.16 acres. On 18-acre land, the shares of vegetables and cotton under Kharif are 16.75 and 1.25 respectively.

(ii) On 8-acre land, the area under Kharif is devoted to cotton (A a/s 100).

(iii) Beyond 18 acres, no cotton is produced; the whole of Kharif land is devoted to vegetables.

(iv) Up to 20 acres, the whole of Rabi land is devoted to wheat (mix 100); no other Rabi crop is produced and the wheat acreage remains steady beyond 20 acres.

In all the changes in the cropping pattern of Farm No. 3, the crucial role is played by the consideration of saving labour at critical times. For instance, vegetables competing with cotton on Kharif land require much less labour than cotton at critical times, e.g. the labour coefficient of vegetables (fertilised) is 1/5th of that of the American cotton (fertilised) in the month of October.

The fact that the Farm No. 3 can *double-crop 20 acres* of irrigated land with 95 M.W.D., while Farm No. 5 can *double-crop only 12 acres* of land with 105 M.W.D. can be explained by the inclusion of vegetables only in the farmer's cropping pattern; vegetables, as we have seen, offer a low labour-content substitute for cotton in the Kharif season.

IV. The increase in income to fixed factors:

(i) As the size of land-holding goes up, the increase in income to fixed factors slackens, declining from an increase of 10.5 per cent between 8 and 9 acres, and 7.1 per cent between 11 and 12 acres; 2.0 per cent. between 19 and 20 acres; 15 per cent between 24 and 25 acres.

(ii) The rate of increase in income is much higher when the farmer is changing from the existing indigenous process to new process, based on fertilisers (see Chapter VI).[2]

VII: 2 CONCLUSIONS

The conclusions reached on the basis of the present investigation are as follows:

First, land, though aggregatively in short supply, is not a 'limiting' factor for all. For Farm No. 5, in our case, labour, and certainly not land, appears to be the limiting factor, the actual size of the holding being 18.50 acres. On the other hand, land is the 'limiting' factor for Farm No. 3 with 12 acres of land, since at this level the existing labour supply is not exhausted in the optimum programme in the crucial months of the year. Hence, given extra land, Farm No. 3 can utilise it with the given labour supply.

Given the use of new capital input (e.g. chemical fertiliser), the Farm No. 5 could use 12 acres of irrigated land and Farm No. 3 could use 20 acres of irrigated land fully, with the given labour supply. Farm No. 5 has, therefore, more land than could be used, and Farm

[2]The percentage increase in income to fixed factors declines as the intensity of land use declines with a rise in the size of holding and low-valued crops replace high-valued ones, e.g. vegetables in place of cotton.

No. 3 has less land than could be used in actual practice. Secondly, the extent to which land could be used with the given labour supply on family farms depends on the nature of the feasible crop alternatives[3]. For instance, Farm No. 3 with a total supply of 95 M.W.D. of labour could use 20 acres of irrigated land fully, because its crop alternatives include 'vegetables' which require much less labour than the substitutable crop, cotton, in the Kharif season. Farm No. 5 with a total supply of 105 M.W.D. of labour could use only 12 acres of land fully, because no low labour-content crop is included in its existing cropping pattern unlike Farm No. 3.

Thirdly, the 'optimum scale' as defined by the size of land-holding is not independent of the choice of techniques in agriculture. When the choice of techniques is mainly oriented towards saving labour, the optimum scale seems to imply bigger size which allows the economic use of machines like tractors and threshing machines. On the contrary, smaller holdings appear to be optimum[4] when the choice of techniques is oriented towards the intensive and the full utilisation of land by applying more labour and intermediary and capital inputs like chemical fertilisers.

The above conclusions are, however, subject to the following limitations. First, we have ignored the role of hired 'casual' labour in farm planning. The decision to appoint 'casual' labour on the part of the individual decision-maker, however, depends on such economic considerations as the certainty of getting labourers at critical times. Because the producer concerned does not have any job to offer at other times, e.g. in the months of October and November. The existence of 17.6 million agricultural labour families, in the aggregate, does not necessarily assure an indivi-

[3]In the existing farming system in India, the labour requirement of crops per acre varies. In the Punjab, for instance, the number of man-days per acre of irrigated wheat is 27.6, of irrigated wheat-gram mixture 19.6, American cotton (irrigated) is 31.0 and that of Desi cotton is 32.8. The labour requirements of crops like rice, sugarcane (Ratoon or planted), jute, etc. are higher than those of wheat or cotton.

[4]This explains the success of a similar technical change in Japan through individually taken decisions in the early periods of economic growth; the average size of farm in Japan was one hectare, i.e 2½ acres. Japan did not introduce any ceiling on land but allowed full freedom to landowners to deal with the tenants.

dual producer in a particular village, at a particular point of time, the availability of casual workers.

The hiring of 'casual' labour to any significant extent requires adjustments in the working pattern of a family farm based on family labour or its combination with the yearly hired labour force, since casual labour has to be supervised and fitted into the working pattern.

Moreover, the decision to appoint casual labour will be determined by the ratio between the wage cost and the expected return from such employment. A low-average wage-rate may be associated with relatively higher rates at critical times, because there is likely to be simultaneous demand for casual labour on bigger farms. Because, 'casual' labour has greater opportunities of self-employment during this period and this raises the supply price.

The existing wage-rate (average daily) in the Punjab for male labour is Rs. 1.73, and for female Rs. 1.32; the daily average wage-rate for harvesting is Rs. 2.49 for male, and Rs. 2.15 for female.

After the introduction of chemical fertilisers, the marginal value productivity of October labour[5] varies between Rs. 10.23 and Rs. 15.1 on Farm No. 5. So long as the wage-rate of casual labour remains low, casual employment can be expected to take place on this farm enabling the producer to utilise more land. Pressures on wage-rate of casual labour are likely to arise in the region concerned because of (i) the spread of technological innovations which tend to increase the simultaneous demand for casual labour and (ii) social pressures may arise from the precarious conditions of such labour families having a short seasonal employment.[6]

[5] *See the Table* ; M.V.P. *of Labour.*

[6] Agricultural labour families belong to the bottom strata of the consumption ladder (according to expenditure) with near-zero expenditure on education, medicine and medical services (National Sample Survey, 9th Round). In a detailed study, Mukherjee has concluded that 75% of the debt of such families in the Punjab is for consumption purposes. See P. K. Mukherjee "A Note on Agricultural Unemployment and Level of Wages," in *Seminar on Rationale of Regional Variations in Agrarian Structure of India*, Mathesan, October 1956, p. 72; also, *Agricultural Labour Enquiry, 1956-57*, Government of India, Delhi.

M. V. P. OF LABOUR

(*rupees*)

Farm No. 5				*Farm No. 3*			
Land Size (acres)	*July L.*	*Oct. L.*	*Sept. L.*	*Land Size (acres)*	*June L.*	*Oct. L.*	*Nov. L.*
12	2.1	15.1	0	8	8.50	0	0
13	8.1	16.2	0	9	,,	0	0
14	1.35	12.4	13.4	10	,,	0	0
15	1.35	12.4	,,	11	,,	0	0
16	,,	,,	,,	12	,,	0	0
17	,,	,,	,,	13	,,	0	0
18	,,	10.22	17.03	14	,,	8.0	0
19	,,	10.23	,,	15	0	17.0	0
20	,,	,,	,,	16	0	,,	0
21	,,	,,	,,	17	0	,,	0
22	0	,,	,,	18	0	,,	0
23	0	,,	,,	19	0	20.33	0
24	0	,,	,,	20	0	40.31	18.16
25	0	,,	,,	21	0	,,	,,
				22	0	,,	,,
				23	0	,,	,,
				24	0	,,	,,
				25	0	,,	,,

The actual amount of casual labour employed on an individual farm in the last instance will depend on the ratio between marginal cost of hiring labour and its marginal productivity.

Additional labour will be employed up to the point where the marginal cost of employing labour (i.e. money wage-rate) is equal to the marginal revenue product of labour ($\equiv$ marginal physical product × price of product). This tendency appears to have been revealed in the studies in Farm Management in different states in India.[7]

On the basis of the above studies, the following conclusions might be drawn:

[7] *Studies in the Economics of Farm Management*, The Directorate of Economics and Statistics, Ministry of Food and Agriculture, States of Punjab, U.P., Bihar, Bombay, etc., 1955-56.

(*a*) the proportion of 'hired' labour as opposed to family labour in the total labour-time, utilised in crop production, increases directly with a rise in the size of farm, the proportion generally varying between 20 per cent and 30 per cent of the total labour employed (in some cases higher, e.g. 80 per cent in Orissa). In the case of Punjab, where the investigated farms are situated, the proportion of hired labour to total labour goes up from 13.1 per cent in the size-group of 5.01-10-00 acres, to 20.5 per cent. in the size-group of 15.01-20.00 acres, and to 64.7 per cent in the size-group of 50.01 and above.[8]

No doubt, the actual amount of 'hired' labour used in a particular case depends also on factors such as the nature of crops produced and the possibilities of double cropping. The relationship between a rise in land size and the increase in the amount of 'hired' labour is well established in the case of all the other states as well.

(*b*) The total labour-time applied per unit of land moves inversely with an increase in the size of holding. This, incidentally, explains why the total cost of labour, when family labour is imputed at the ruling market wage-rate, moves directly with a fall in the size of holding.

(*c*) The productivity per unit of land decreases with the size of holding as the intensity of cropping declines. This trend in the gross output per acre is observed, more or less, practically in all regions studied.

The above conclusions of the farm management surveys imply that the amount of an input to be applied (labour in the present case) in the present context is determined by basic economic calculations of cost and return. The small producers apply more labour per unit of land, since 'family labour' is a 'free' 'fixed factor' in the short period. Given the low rate of substitute between leisure and income of low income group small farmers and the lack of alternative employment opportunities, family labour has no 'opportunity' cost. In such cases, the maximum utilisation of the available land coincides with the maximum of family income.

[8] "Hired Labour Use in Crop Production," *Agricultural Situation in India*, January 1962, Vol. XVI.

On larger farms, the application of less labour could be explained in terms of (i) the change of leisure-income preference of family members (the absence of any female labour on the largest investigated farm may be ascribed to it), and (ii) the amount of hired labour being determined by cost and return calculations. In such cases, the point of the maximum family income does not necessarily imply the full utilisation of land as an input.

It might be argued that the labour coefficients used in our programmes have ignored the possibility of increasing returns to scale, i.e. the labour requirements increasing less proportionately than a rise in the size of holding. For instance, a farmer is likely to require less than double labour-time in ploughing two acres of land than in one acre.

The above possibility, however, would not invalidate the logic of our argument, namely, that the full utilisation of land as an input might not be an economic proposition for all the individual producers concerned. Secondly, it must be acknowledged that the labour constraints in our models have ignored the possibility of fixed investments in labour-aiding devices which might increase the effectiveness of the present labour supply.

The availability of suitable labour-saving devices in India is, however, limited. (a) Because the aggregative land-labour ratios emphasize the urgency of producing land-saving rather than labour-saving devices, (b) the urgency of raising yield of crops per acre of land emphasizes the allocation of resources to the production of inputs, such as fertiliser, seeds, rather than to tractors (cf. Introduction).

Moreover, the labour-saving devices on large farms will be effective only if the pattern of the leisure-income substitution of these richer families remains unaffected, or if the labour thus released does not take up non-farm jobs. Such redistribution of labour-time between farm and non-farm jobs is a possibility, since such diversification of the occupational pattern is present today among the big farmers.[9]

[9]A pilot survey of the fourteen villages in the U.P. and the Punjab concludes that in the four villages of Hissar (Punjab), the most important combination of occupation is agriculture and trading and services; in the two villages

In principle, however, it is likely that the effective labour supply on the bigger farms like Farm No. 5 might increase in the short period due to a change in marginal rate of substitution of leisure for income even if no labour-saving devices are introduced.

Such changes may arise from factors such as (i) new 'wants' being created by the developing industries, (ii) social approval for hard work rather than for leisure, (iii) individual initiative in breaking social restrictions, if any, on field work (large farmers in India often belong to the Brahmin caste with injunctions against field work), (iv) taxation poliy of the state, e.g. the imposition of a high Land Tax.[10]

Lastly, it might be pointed out against our conclusions that the present study is based on two crucial assumptions, (*a*) that chemical fertiliser is used; this has inflated the labour coefficients for the crop alternatives, (b) that double-cropping is allowed ; this has made the use of land in both the seasons necessary, if land is to be fully utilised.

The justification of the above assumptions is as follows: First, we are interested in finding out how agricultural output and income might be raised in India (see Introduction). Hence, we are concerned with what might happen if new technological possibilities become feasible in the region concerned. Secondly, since water, soil and manurial conditions allow double cropping with major crops in the region concerned, the effective supply of land increases.[11] This possibility might not be present in other regions.

in Rohtak, farming and rent-receiving or farming and labour applies to small farmers; an economic survey of Shamaspur village shows that 14.9% of the households combine two operations, e.g. farming and trading; farming and rent-receiving, and farming and interest-receiving. The largest cultivating family in the village, owning 1/3 of the entire land, combines farming (with hired labour) and trading and money-lending and rent-receiving. A similar trend was witnessed in the past. See *A Pilot Survey of the Fourteen Villages in the U. P. and the Punjab*, Asia Publication, 1959. *An Economic Survey of the Village of Shamaspur*, Asia Publication, 1959. *The Food Grains Enquiry Committee Report*, Government of India, 1959, specially, pp. 410-11 and 446.

[10] In Meiji, Japan, the imposition of a highland tax forced landowners to switch over to technical changes in the short period.

[11] In the exisiting system as well, farmers have double-cropping of land (cf. Chapter VI).

Thus, we find that the possibilities of utilising land fully (cf. Introduction) on individual farm units depend on factors such as the availability of family labour and the nature of the crop alternatives open. There is no reason to assume, *a priori*, that all agricultural producers in India would use their land fully in their own farm planning.

So far as land is the only limiting factor, an individual decision-maker might use his land supply fully; in such cases, the individual and social net products coincide in the given situation in India. On farm holdings, where land is not the only limiting factor, the extent to which land is utilised depends on the economic calculations of the individual involving a large number of factors;[12] in such cases, the individual and the social marginal net product do not necessarily coincide. Two alternatives are there. First, the full utilisation of the available land might be given up as an independent objective of the macro-economic planning (cf. Introduction); secondly, keeping the macro-economic objective in view, special measures (e.g. legal, social) might be taken to create a situation where individual decision-makers will use land fully on their individual holdnigs, although, in the absence of such measures they may not do so. (see Chapter VIII).

[12]An interesting line of investigation is to re-examine the present programmes after introducing 'hired' labour as an activity in the models to find out what happens when we allow for the hiring of casual labour. The author has to postpone this line of investigation due to the lack of resources.

CHAPTER VIII
CONCLUSIONS

IN THE first section of the present chapter, we point out some of the main significant contributions of our study as well as its principal limitations. This is followed by a discussion of the policy recommendations which arise from our previous discussion.

VIII : 1

Our study has in the first place sought to throw new light on the *raison d'etre* of the family farming system in India—more specifically on (i) the rationale of its operation in the past, e.g. existing technique (Chapter III), the choice of cropping pattern (Chapter IV), the response to uncertain situations (Chapter II), and the response to land-saving innovations (Chapter V); (ii) its potential future role, e.g. the importance of small family farms with plenty of family labour for the successful adoption of land-saving technological innovations such as chemical fertiliser and hybrid seeds (Chapters VI and VII) has already been noted.

Secondly, the conclusion reached in the present study, namely, that the agricultural producers in India raise their family income as modified by certain considerations arising from the environment has implications for policy. Thirdly, the present study, by its very approach, points to the importance of carrying out further sample studies in different regions in India. To test our hypothesis further, we need such empirical investigations specially designed for the purpose. Fourthly, the present study emphasises the importance of relating the macro-economic objective of raising agricultural output to individual farm-planning.

As regards the major limitations, the present study, in the first place, suffers from the limitations of the available statistical data. For instance, we could not examine the relationship between expected income and income variability in the case of the fertilised processes as we did in the case of the non-fertilised processes of some crop activities simply because we do not have the necessary time series data of yield (cf. Chapter IV). Again, the availability of input-output coefficients, covering activities such as the 'livestock,' would have increased the realism of the alternative farm plans formulated for individual farms (Chapter VI). Some fruitful lines of investigation were, thus, not pursued due to the lack of data.

Secondly, the shortage of time and resources prevented us from extending our study over more than one region. We settled on Punjab. Punjab is thought to offer great possibilities of technical change on account of favourable soil and climate, irrigation facilities, and a relatively favourable land-labour ratio. Since previous efforts at introducing innovations were largely concentrated on this region, we have also more empirical data than what is offered by other regions. This regional bias makes us wary of any generalisation as to the whole of India.

VIII: 2 The Policy Implications

In our introduction to the present study, we pointed out two important objectives of the agricultural planning in the agricultural sector in India. First, agricultural output must be raised by raising the yield per unit of land through the introduction of innovations such as irrigation facilities, chemical fertilisers, hybrid seeds, etc. Secondly, employment opportunities in agriculture must be created by utilising the available supply of land fully, given the population pressures and the non-availability of fresh land.

Policy recommendations could be grouped into two parts. Group A includes policy recommendations oriented towards the first objective. Group B includes policy recommendations oriented towards the second objective.

Group A

First, in Chapter IV we saw that given the relationship between the expected income and income variability in the existing system (i.e. the shape of the EV curve), and the practical possibilities of raising income, from the individual's standpoint, are limited.

Over time, the spread of scientific cultivation incorporating new inputs and new knowledge could change the shape of the EV curve by reducing the income variability and raising the expected income.

In the short period, however, what is technically suitable might not be economically feasible for an individual producer. New resources and new knowledge might not reach the farm gate. Even if they are technically accessible, the lack of credit facilities might hamper his ability to adopt changes. A defective marketing system could destroy his 'willingness' to change even if he is 'able' to do so. The availability of one input without crucial complementary inputs might increase the risk element in existing situations rather than diminishing it. For instance, the effective use of chemical fertiliser not only requires, as an essential precondition, an adequate supply of water, but also the ability of the decision-maker to undertake actions in many directions, e.g. managing bumper crops at harvest time.

Individual producers require advice and guidance in periods of rapid technical changes. Such advice and guidance, in the context of India, depends on the efficiency of the extension services; the latter, in turn, depends on the following factors.

(*i*) *Research and New Knowledge.* Since the marginal social net product from investments in research[1] tends to be much higher than the marginal private net produce, public authorities have a clear role to play in fostering research. It has been rightly observed, "an extension service can only be as good as the research information available to it."[2]

(*ii*) *The Embodiment of New Learning in Capital Inputs.* The extension officers could take scientific knowledge to the producers when such knowledge is embodied in capital inputs[3]—whether as tractors or improved seeds. The supply of new inputs at stable

[1]Cf. Professor Solow's discussion on this point. R.M. Solow, *Capital Theory and the Rate of Return*, Amsterdam, 1963, p.65.

[2]Australian paper (C/393) in *United Nations, Science and Technology for Development*, III. *Agriculture*, New York, 1963, p. 189.

[3]An example of disembodied investment where new knowledge need not be contained in new capital inputs is a change in the time of sowing.

prices will create new technical choice variables for decision-makers in agriculture.

(*iii*) *The Extension Services and Bridging the Gap between Research Centres and Individual Farmers.* The efficient structure of the Extension Services is an important condition for speeding up the process of technical change in agriculture. The tasks to be performed by the extension officers appear to be twofold: first, they are expected to disseminate new knowledge of science and technology among the individual producers. Secondly, they are expected to act as a liaison between the research centres and the actual needs of the producers.[4]

(*iv*) *Producers' Organisations and the Extension Services.* Producers' organisations, in a small farming system, could provide a liaison between the extension officers and the actual decision-makers in agriculture. Through such organisations, new knowledge and inputs could be made familiar to small producers. Moreover, such organisations could act as the centre of education through group discussions and demonstrations. In Japan, producers' organisations in the villages played a crucial role in the rapid spread of innovations in a small farming system. Secondly, the speed of technical change could be facilitated through administrative measure, mitigating risk elements in the farm planning environment (cf. Chapter II).

(*v*) *Price policy and risks arising from demand fluctuations.* A policy to keep prices stable—both of the products and inputs—will contribute greatly to the reduction of risk elements.

[4]The special advantages of supervision and evaluation of field trials and demonstrations by research scientists as experienced in the United States are: "First they take research scientists into the field where they come into direct contact with production problems, members of the agricultural advisory staff and practising farmers. This usually proves useful all round, particularly in providing for the kind of 'feed back' to central research stations and laboratories that is necessary if successive road-blocks to increased production are to be systematically attacked as they appear. Second, observation of local field trials and test demonstrations by farmers can play an important role in their developing a rational scientific attitude towards farming. In observing local testing and field trials, farmers may learn the basic steps of observation, measurement, comparison of results and drawing of conclusions that are the fundamentals of rational thinking and science-based agriculture"—*U.S.A. Paper* (C. 1965). See United Nations, *Science and Technology*, 1963, pp. 189-90.

(*vi*) *Future land structure.* A clear choice as to the future land structure in India and the setting up of a legal machinery, consistent with the choice made, would reduce risk elements from the environment. At present, the alternatives include small owner-cultivators, large commercialised private farms, cooperative joint farms and large state farms. An individual's decision to invest and to undertake improvements might get postponed when he is kept ignorant of what is going to happen in the land system. Moreover, technical change in agriculture is affected by the interaction of many individual factors. For instance, the creation of small owner-cultivators is likely to create a demand for a series of complementary measures; the public authorities have to undertake new responsibilities. On the contrary, the creation of big commercialised farms might create a demand for alternative provisions for the tenants who are to be ejected from land.

(*vii*) *Insurance against risk arising through changes.* Innovating producers' initial fears in handling a new input, a new method, could be overcome by offering special compensations against any possible hazards involved. For instance, one or two years' tax exemption for an innovating producer, or a system of deferred payments for expenses might increase the spread of innovations; producers are thereby enabled to meet eventualities involved in the initial experiments with new capital inputs. In this context, the possibilities of insuring crop activities based on *new processes* (e.g. fertilised processes of major crops) could be explored.

Thirdly, in the context of the alternative farm plans based on new capital inputs—epitomised in chemical fertilisers—we found that the credit need created by new farm programmes might not be met efficiently by the existing system of credit.

For a rapid spread of innovations, credit, in India, must be accompanied by knowledge and training. Moreover, in the Indian context, an integrated approach to credit appears to be the only way of overcoming the limitation because of the following factors.

First, the problems of credit and marketing in India are now closely linked up. Small farmers invariably need credit accommodation to tide over family distress caused by crop failures and the death of cattle. Credit for consumption purposes are not met by the organised credit structure. Private money-lenders, supplying such consumption credit, easily secure control over the small farmers' marketable surplus. Unless small farmers are assured of credit for their legitimate consumption needs, the hold of private money-lenders will remain. Only by integrating credit with marketing can public policies assure an individual decision-maker of a regular credit accommodation as well as an opportunity of receiving the full price of his own product.

Secondly, when credit and marketing arrangements are integrated, it is easier to provide loans on the basis of crop security. This would help small farmers with few tangible assets to obtain financial accommodation, otherwise inaccessible to them.

Thirdly, through such integrated credit-cum-marketing societies, new capital inputs can be supplied to producers as 'loans.' Since loans in such real terms are practical only when there is some supervision as to their proper use, the integrated credit-cum-marketing societies can act as the institutional basis for the spread of such supervised 'real' loans.[5]

The present investigator feels that the success of the spread of innovations in the countryside will largely depend on the growth of producers' organisations-cum-marketing-trading societies. The growth of such societies in the short period will largely depend on the ability and the willingness of the public authorities to take up the responsibilities involved in the supply of new resources and knowledge, the supply of credit and the arrangements for marketing.[6]

Agricultural output in India cannot be expected to rise automatically in the short period. The ability of the small family farmers, to overcome the environmental limits will largely depend on the efforts of the public authorities at removing such constraints (see Chapters II and V). There is no short cut to the development of an agriculture based on science and technology.

[5]The instances of the wastage of irrigation facilities and seeds or fertilisers are likely to be more in a system, where thousands of individuals are concerned with their use, than when a few large farmers are responsible for their use.

[6]Small family farms, however efficient from production standpoint with labour-intensive techniques, are in a poor bargaining position vis-a-vis the market.

In conclusion, it must be pointed out that the above policy implications are deduced from the conclusions of our study, i.e. the limitations on small family farmers aspiring to raise farm income and productivity arise mainly from the environmental factors and not from any supposed absence of economic motivation. If our hypothesis is refuted through further empirical tests, the policy implications are likely to be different, the emphasis shifting to a basic change in the institutional structure, e.g. the creation of commercialised large private farms or the creation of large state farms.[7]

Group B

When the full utilisation of the existing land supply is accepted as a secondary objective of public policies (e.g. on account of employment, foreign exchange or social considerations), special policy measures are needed. Such special policy measures are discussed below.

First, the ceilings on large landholdings. A ceiling may be imposed on large landholdings to reduce their size to the optimum level, i.e. the level at which the land is fully utilised by the decision-maker with the available supply of labour (see Chapter VII).

The discussion on 'ceilings' on landholdings has so far been carried on in India, largely in terms of some abstract concepts of equality and justice (see Introduction). Thus, according to the Second Five Year Plan, "Reduction of disparities in the ownership of land is also essential for developing a cooperative rural economy, for cooperation thrives best in homogeneous groups in which there are no large inequalities."

In terms of the present discussion, the imposition of ceilings on landholdings appears to have an immediate economic implication, i.e. the creation of optimum-size farms. In determining the optimum-size holdings, various factors have to be taken into account, e.g. the existing land-labour ratio, the possibilities of

[7]Such a basic change in the institutional structure might still be regarded necessary in an economically backward country, simply as a means to reduce the time-element necessarily involved in the spread of technical changes through the independent decisions of the multitude of small farmers.

double-cropping, the nature of the crops produced in a region and their labour requirements, etc.

From our study of a single region, no generalisation is, of course, possible as to the level of ceiling in other parts of the country. For the state of Punjab, however, the imposition of a ceiling on existing holdings at 30 standard acres[8] is devoid of any economic significance, since it is making a very large farm (in the Indian context) a bit smaller without creating the optimum-size holdings where land will be utilised fully by individual producers motivated by their own economic interests (see Chapter VII).

Objections may be raised against a policy of land redistribution on political, ethical or economic grounds (i.e. large farms enjoy advantages of scale).

A second policy alternative may be the spread of tenancy. Since the 'operating unit' need not necessarily be the 'ownership unit,' the objective of creating the optimum-size holdings may be secured by distributing the excess land on large farms among tenant-farmers. For small tenant-farmers, the point of the maximum family income may coincide with the point of the full utilisation of land.

Against the above possible policy recommendation, the following objections may arise:

First, when the output derived from a small-size holding is divided between a tenant operator and an owner, the actual producer often lacks financial resources to adopt new technical changes. The saving investment calculations really enter into the farm planning environment, only after the survival requirements of a family are met.

Secondly, the tenancy system in India, is already associated with many undesirable features, all joining against the incentives of

[8]Ceilings on landholdings differ from state to state: (i) Punjab: 30 standard acres and not exceeding 60 ordinary ones (a standard acre means one acre of perennially irrigated land, i.e. land with water throughout the year, or 2 acres of seasonally irrigated land or 3 acres of dry crop land). (ii) Andhra Pradesh: $4\frac{1}{2}$ times the family holding, i.e. 27 to 324 acres (a family holding varies from 6 acres of wet land to 72 acres of dry land and one extra family holding is allowed for every member in excess of 5 acres).

small tenant farmers to adopt innovations to raise family income, e.g. the disproportionately small share of the tenant into the fruits of his own efforts. At best, the prospect in India in future is a combination of a "sellers' market" in tenancy in the absence of alternative regular occupations for landless rural families (e.g. 30 million in 1955) and for the new additions to labour force due to the rise in the population. The economic implications of the weakening bargaining position of tenants vis-a-vis owners are unlikely to be avoided, simply by legislative measures or fact-finding commissions. Hence, further spread of tenancy is undesirable for social stability and social cohesion.

A third alternative lies in the administrative measures. The economic results of the creation of the optimum-size land-holdings can be approximated by the imposition of administrative controls on bigger holdings, e.g. a penalty tax on the non-utilisation of land along with a simultaneous system of 'bonuses' for the full utilisation of land.

The possible objection arising against the above policy recommendation may be that such administrative controls will hamper the 'incentive' and the 'initiative' of big farmers. Moreover, since the basic divergences between the private and social net products on big farms are not removed by such control measures, there will remain a permanent source of friction between the public authorities and the individuals.

Lastly, the public authorities might give up the consideration of the full utilisation of land available as an independent objective of agricultural policy.[9]

Possible objections arising against the above policy recommendation are. First, the full utilisation of land as a scarce resource is needed, from the standpoint of society, to meet food shortages, to create new employment opportunities for the unemployed; in

[9] Since the top 5% of the rural families, having more than 20 acres, own 42% of the cultivated land, it appears to be of extremely limited significance to the present investigator in calculating the warranted rate of growth in agriculture, in terms of the total available land, total available labour force and the effectiveness of intermediary capital inputs in relation to the total available fixed resources, i.e. land and labour.

short, to deal with the problem of rural poverty and low productivity. Secondly, when land remains unutilised on large holdings in a village, while there are landless agricultural families without any alternative source of regular employment, it acts as a permanent source of social and political frictions in the rural areas. Such frictions often appear in India, in the form of 'caste' frictions affecting social cohesion and stability. For instance, the agricultural landless families in India, tend to belong to lower castes, while many of the very large land owners often belong to the highest caste, viz. the Brahmins.

Since the actual choice of any particular policy recommendation, as discussed above, will be determined not merely on economic grounds, the present investigator has refrained from giving any unique solution to the problem.

To conclude, the present study has treated policy recommendations as a 'technical ought,' i.e. what the public authorities ought to do if they want to fulfil certain objectives, generally accepted in India as the goals of macro-economic planning.

Three steps are involved here: first, certain objectives considered 'desirable' are formulated (cf. Introduction). Second, a model has been built up as regards the behaviour pattern of the agricultural producers (cf. Chapters II, III, IV and V). Third, we recommend certain acts of policy, which, if the model is consistent with the reality, might help us to attain the desired objectives. Lastly, we assume that the policy recommendations as discussed here are the most efficient in the given context—efficiency being defined as the capacity to achieve desired objectives while creating as little 'undesirable' effects as possible. For instance, a recommended policy creating large private farms through a modern enclosure movement might speed up the spread of the scientific cultivation in Indian agriculture. Large private farms will enjoy better financial resources, better bargaining position in the market and greater efficiency in the use of resources due to the advantages of scale. In India, its undesirable effects, however, would be that (i) millions of small farmers uprooted from the land could not be offered any alternative employment opportunities at the current stage of the economic development of the country; (ii) the possibilities of land remaining unutilised on large farm holdings would increase.

If the behaviouristic model underlying the study is consistent with the reality, the most efficient way of raising agricultural output in India appears to be to make the small family farming system viable through the persistent efforts of the public authorities.

APPENDIX A

WAGE COST OF LABOUR AND COST CALCULATIONS OF INNOVATIONS

I. Weed, Pest, Diseases

Co-ordinated research schemes on weed control were in operation at the National Chemical Laboratories, Poona (from 1953 to 1957 for the local manufacture of pure weedicides), at the Bose Institute, Calcutta (from 1952 to 1958), in Madras (at Coimbatore from 1952 to 1956), for evolving proper weedicidal schedules. Following is a summary of the Reports of the first batch and current experiments:

(*i*) *Wheat:* Experiments at Chhabra (Himachal Pradesh) showed that hand-weeding gave a maximum net profit of Rs. 11/12 per acre, while treatment with sodium salt of 2, 4-D gave a net profit of Rs. 6.94 per acre. The other weedicide treatments resulted in financial loss.

(*ii*) *Jowar:* Hand-weeding gave 10 times more yield than unweeded control, while weedicide-treated plots gave 5 times more yield only in Khari jowar (at Nagpur).

(*iii*) *Maize:* Decrease in yield over control when weedicides were applied (Himachal Pradesh). Hand-weeding gave about 40% increased yield over control.

(*iv*) *Paddy:* At the Bose Institute, Calcutta, yield of the unweeded paddy plots is the lowest, while that of the hand-weeded plots is the highest (weeded six times during the season).

The present cost of artificial weedicide varies from Rs. 68.50 to Rs. 63 per acre; the average daily wage of hired labour Rs. 1.62 as per day (male) and Rs. 0.68 (female). (*See Annual Report of the Indian Council of Agricultural Research for 1957-58.*)

II A Specific Case of the Improved Tillage of Paddy—its Requirements of new Capital Inputs and Labour and Analysis of the Cost Items:

APPENDIX A

1	2	3	4
Method of Cultivation	*Gross Income (Rs)*	*Expenditure (Rs)*	*Net Income (Rs)*
1. Broadcast (local method)	153.00	71.69	81.32
2. Transplanted (local method)	405.00	148.00	256.94
3. Japanese method	1013.50	314.25	(1) with f.y.m. 699.25 (2) with green manure 760.50

Items	*Broadcast*	*Transplanted*	*Japanese*
1. Ploughing	18.00	36.75	37.19
2. Planking	4.50	4.56	4.62
3. Bunding	1.50	3.56	1.50
4. Seeds and sowing	16.00	4.75	15.90
5. Transplanting	nil	10.50	13.50
6. Manuring (f.y.m.)	nil	10.00	106.50*
7. Green manuring	nil	nil	40.25**
8. Chemical manuring	nil	nil	79.25
9. Irrigation	15.50	16.94	17.00
10. Thinning	3.00	nil	nil
11. Interculture	nil	31.50	6.69
12. Harvesting, threshing, clearing	10.19	27.00	28.94
13. Land revenue	3.00	3.00	3.00
Total	71.69	148.00	314.25

*Rs. 40.25 being the cost of green manure is a substitute of farmyard manure costing Rs. 101.50; labour cost calculated at market wage-rate.

**The analysis of the individual cost items shows that numbers 1-5, 9, 10, 11, 12, 13 do not pose any real problem so far, as the adoption of the Japanese method of rice cultivation in India is concerned. The real input needed is labour which is in plenty of excess supply in the rice-producing regions in India. Moreover, many of the new tasks are spread over one cycle of crop production, and thus imply a rise in the working time of the cultivators in slack working periods. Items 6 and 7 require a lot of organisation at village level but no expenditure of foreign exchange. Item 8, i.e. chemical manuring, implies the expenditure of foreign exchange either for the import of fertiliser plant or for the direct imports of this input.

In the rice-growing tracts in India, the population pressure on land is reflected in the low wage-rate of hired labour; for example, the daily wage-rates of 'hired labour' in Orissa, where family farms have plenty of surplus labour, are Rs. 0.72 (male) and Rs. 0.49 (female) against all-India averages of Rs. 1.90 and Rs. 0.68 respectively. In such regions, the extension officers have found easy response of the producers to the Japanese method; the limiting factor appears to be the non-availability of fertiliser in the region. This, in turn, is consistent with our discussion in Chapter V.

APPENDIX B

LINEAR PROGRAMMING MODELS AND DESCRIPTION OF FARMS CONCERNED

ABBREVIATIONS USED IN OUR MODELS

(*a*) *Commodities:*

Cotton A	American Cotton (long staple)
Cotton D	Desi Cotton (short staple)
W-G	Wheat-Gram mixtures
S	Sarson
St	Cotton sticks
B	Bhusa (dried fodder)

(*b*) *Fertilisers and Mannres:*

F.Y.M.	Farmyard Manure
A/S	Ammonia Sulphate
Mix.	Fertilisers consisting of two or more chemical elements
Phos.	Phosphates
Amount	100 *A* Rs. 20 worth 100 *S* mix Rs. 25 worth

(*c*) *General:*

M.W.D.	Man Work Day (8 hours a day and 25 days a month)

(*d*) *Seasons*

K—Kharif (Monsoon)	Principal Crops: Jowar, Maize, Rice, Cotton, Sesamum, fodder vegetables, Rape, Bajra, Toria, etc.
R—Rabi (Winter) Sugar cane: two-seasons crop	Principal crops: Wheat, Gram, Barley, Oats, Linseed, vegetables, fodder, etc.

General Features of Our Models

(1) In both the cases of Farm No. 1 and Farm No. 5, we have used the fertilised and the indigenous processes of crops. In other cases, we have concentrated mainly on fertilised processes.

(2) The use of chemical fertiliser raises labour requirement—specially in weeding and harvesting operations; however, its use also saves labour when labour is used for collecting farmyard manure in the indigenous processes.

(3) We have allowed the possibility of double-cropping land with two major crops, e.g. cotton and wheat, because (i) fertiliser takes care of the considerations of soil fertility, and (ii) fertiliser reduces the length of time required for cotton.

(4) We have used labour constraints as we find it reasonable; we have taken into account the following factors; the use of fertiliser raising labour requirement, types of the seasons concerned, nature of the crop concerned. We have varied months used in the case of the farms concerned to investigate what happens to surplus labour in the slack months (i.e. not sowing or harvesting time).

FARM No. 1a (BASIC MATRIX)

Resources	*Avail*	*Unit*	*Cotton* (*D, K*) 5c*F.Y.M*,	*Cotton* $A, \frac{a}{s}-100$	*Fodder* (5c*F.Y.M*).	*Fodder* $\frac{a}{s}-100$	*Maize*	*Maize* (*mix* 100)	*Sugar cane* (*F.Y.M.*)	*Sugar cane* $\frac{a}{s}-75$	*Gram* (*mix* 100)	*Gram*	*Wheat* (*F.Y.M.* 100 *mix*)	*Wheat* (130 *mix*)	*Wheat Gram*	*Wheat Gram* (*mix* 100)
Z-c	(*Rs.*)		—57.30	—184.20	—94.90	—124.90	—57.90	—59.90	—59.30	—82.40	—34.90	—20.90	—170.60	—176.60	124.30	—164.0
Labour July	95	MWD	6.20	8	8.50	2	9.00	9	5.90	6.66	..	..	2.5	1.10	2.20	2
Labour August	95	MWD	4.00	4	7.50	4	7.10	8	8.80	10	8	4	2.5	1.70	1.10	3
Labour September	95	,,	5.90	6	6.40	4	6.40	7	11.80	13.3	4	8	2.50	2.80	2.20	3
Labour October	95	,,	9.80	10	3.20	7.30	6.40	6	..		4		2.50	2.80	3.30	1
Land Kharif	8.60	acres	1	1	1	1	1	1	1	1						
Land Rabi	8.60	,,							1	1	1	1	1	1	1	1

Note : (a) We have included both the indigenous and the fertilised processes.

Description of the Farm : (1) Situation: Village Chaudhirwala, Distt. Amritsar, Punjab,

(2) Land Supply : 8.60 acres of irrigated land, i.e. 8.60 acres of Kharif and Rabi land respectively,

(3) Labour supply: 3 adult males working full time and 2 females working part-time (all family members), i.e. 95 MWD in total (i.e. 25+25+25—10+10).

FARM No. 2 (BASIC MATRIX)

Z—c		*Rs.*	—156	—40	—125	—129	—130	—100	—137.5	—100
Resources	*Avail*	*Unit*	*Cotton A* ($\frac{a}{s}$ 100) (*K*)	*Cotton D* ($\frac{a}{s}$ *FYM*) (*K*)	*Maize* (*mix* 100) (*K*)	*Paddy* ($\frac{a}{s}$ 100) (*K*)	*Fodder* ($\frac{a}{s}$—100) (*K*)	*Wheat* (*mix* 100) (*R*)	*Wheat* (*mix* 130) +5*cF.Y.M.*	*Melon* (5*c F.Y.M.*) (*R*)
Labour—June	65	MWD	6	5	5	12	2.7	3	3	6
Labour—July	65	,,	8	10	9	4	7.7	1	1	2
Labour—September	65	,,	6	5	7	6	2	2	3.5	2
Labour—October	65	,,	10	10	6	6	2.3	2.[illegible]	3.3	2
Land—Kharif	9.6	acres	1	1	1	1	1			
Land—Rabi	9.6	,,						1	1	1

Description of the Farm : (1) Situation : Village—Chaudhirwala. Dist. Amritsar , Punjab

(2) Land supply : 9.60 acres of irrigated land (i.e. 9.60 acres of Kharif and Rabi land respectively.

(3) Labour supply : 2 adults working full-time and 2 females working part-time (all family members), total 65 MWD (25+25+10+5).

FARM NO. 3 (BASIC MATRIX)

2—*c*		Rs.	—194	—235	—36	—102	—126	—170	—136
Resources	*Avail*	*Unit*	*Cotton A (a/s 100) (K)*	*Fodder (a/s 100) K+R (b)*	*Bajra (FYM) (K)*	*Pulses (M 100) Phos) (K)*	*Vegetablees (mix 10o) (K)*	*Wheat mix 10o) (R)*	*Wheat Gram Sarson (mix 100)*
Labour—June	95	MWD	6	2.0	5	6	2	3	2.5
Labour—July	95	,,	8	4.0	5	6.4	2	..	5
Labour—September	95	,,	6	2.0	8	4	2	1.7	1.5
Labour—October	95	,,	10	2.0	2	4	2	1.3	1
Labour—November	95	,,	2	8.0	..	3.2	2	2.7	3.8
Land—Kharif	12.5	acres	1	1	1	1	1		
Land—Rabi	12.5	,,		1				1	1

NOTE : We have included the month of November for further investigation into the labour requirement when fertilised processes are used. We have included the indigenous process of Bajra, because it gives a low-labour content crop substitute for cotton. Since this farm was investigated parametrically as well, we deliberately included Bajra in the matrix. (*b*) Covers land in both the seasons.

Description of the Farm : (1) Situation : Village—Daroli Bhaiki, Dist. Ferozepur, Punjab.

(2) Land Supply: 12.50 acres of irrigated land (i.e. 12.50 acres of Kharif and Rabi Land respectively).

(3) Labour supply : 3 adults working full-time and 3 women part-time (all family members) ; total 95 M.W.D. (i.e. 25+25+25+5+5+10).

FARM NO. 4 (BASIC MATRIX)

Z—c		*Rs.*	—194	—125	—133	—405	—170	—183	—140
Resources	*Avail*	*Unit*	*Cotton K A a/s 100*	*Maize K Mix Mix 100*	*Toria K Mix 100*	*Vegetables K + R (a) Mix 149 (a)*	*Fodder K + R (b) Mix 100 (b)*	*Wheat R Mix 100*	*Wheat Gram Mix 100*
Labour—June	110	MWD	6	3	4	12	.5	2.4	.66
Labour—July	,,	,,	8	5	3	11	.5	1.4	.66
Labour—September	,,	,,	6	6	6	11	.5	3.6	3.3
Labour —October	,,	,,	10	0	6	11	7	3.3	3.3
Labour—November	,,	,,	2	0	6	22	8	1.6	3.3
Land—Kharif	13.17	acres	1	1	1	1	1		
Land—Rabi	13.17	,,				1	1	1	1

(*a*) covering land for two seasons
(*b*) covering land for two seasons

Description of the Farm : (1) Situation : Village—Chaudhirwala, District—Amritsar, Punjab
(2) Land supply : 13.17 acres of irrigated land (i.e. 13.17 acres of Kharif and Rabi land respectively).
(3) Labour supply : 3 adult males (including 1 permanently attached worker) and 4 females (family members). Total labour-time is 110 M.W.D. (i.e. 25+25+25+10+10+5).

FARM No. 5 MODEL I (BASIC MATRIX)

Resource	Avail	Unit	Cotton K $A, \frac{a}{s}$-100 Irrigated F.Y.M.	Cotton K $A, \frac{a}{s}$-150 Irrigated F.Y.M.	Cotton K, D, FYM Irrigated	Cotton K $D \frac{a}{s}$-150 Irrigated	Fodder KFYM Irrigated	Fodder $K \frac{a}{s}$ 100 Irrigated	Bajra KFYM Irrigated	Wheat R Unirrigated	Wheat R Irrigated	Wheat R mix.20 irrig.	Wheat gram mix.100R irrg.	Wheat gram R Un-irrig.	Wheat gram R mix 1 50 Irrig.	Wheat gram Sarson Un-irrig.	Wheat gram Sarson mix 47 Un-irrig.	Fodder R Mix 100 irrigated
Z-c		Rs.	-207.66	-215.2	-74.5	-140.5	-96.5	-126.5	-64.5	-85.7	-92.0	-166.5	-102.5	-90.6	-165.6	-70.5	-87.2	-101.5
Labour—July	105	MWD	6.20	3	8	8	9	2	8	1.00	.50	1.5	1	2.70	2	0.85	1.90	(b)
Labour—August	,,	,,	3.90	4	4	4	8	4	7	2.00	.80	2.5	2	1.70	3	1.70	2.80	
Labour—September	,,	,,	5.80	6	6	6	7	4	4	3.50	1.20	3.0	3	2.10	3	1.80	2.80	
Labour—October	,,	,,	9.75	10	10	10	6	7	..	2.95	1.20	1.50	2.5	2.10	1	2.60	2.30	
Land—Kharif (irrig.)	8.50	acres	1	1	1	1	1	1	1									
Land—Rabi (irrig.)	8.50	,,										1	1		1			1
Land—Kharif (unirrig.)	10.25	,,								1 (a)	1 (a)			1 (a)		1 (a)	1	
Land—Rabi (unirrig.)	10.25	,,								1	1			1		1	1	

(a) not allowing for double-cropping on unirrigated land

(b) Rabi fodder does not require labour in the months concerned

FARM NO. 5 MODEL II (BASIC MATRIX)

Z—c		*Rs*	—207.66	—215.20	—74.50	—140.50	—96.5	—126.5	—64.5	—166.50	—102.5	—165.6	—101.5
Resources	*Avail*	*Unit*	*Cotton K a/s100 FYM*	*Cotton K a/s 150A FYM*	*Cotton K D FYM*	*Cotton K Da/s150*	*Fodder K FYM*	*Fodder Ka/s100*	*Bajra K FYM*	*Wheat R mix 100*	*Wheat grain R*	*Wheat gram R mix 150*	*Fodder R mix 150*
Labour July	105	MWD	6.20	8	8	8	9	2	8	1.5	1	2	
Labour August	,,	,,	3.90	4	4	4	8	4	7	2.5	2	3	
Labour Sept.	,,	,,	5.80	6	6	6	7	4	4	3.0	3	3	
Labour October	,,	,,	9.75	10	10	10	6	7		1.50	2.5		
Land Kharif (irrig.)	18.30	acres	1	1	1	1	1	1	1				
Land Rabi (irrig.)	,,	,,								1	1		1

(a) This model is used for the parametric programming as well.
Description of the Farm : (1) Situation : the same
(2) Land supply : total landholding is assumed to be fully irrigated.
(3) Labour supply : as before

Description of the Farm

(1) Situation : Village—Daroli Bhai Ki ; District—Ferozepur, Punjab
(2) Land supply : 18.75 acres total; 8.50 acres of irrigated and 10.25 acres of unirrigated land (i.e. 8.50 acres of Kharif and Rabi land and 10.25 acres of unirrigated land—3 land constraints are used).
(3) Labour supply : 3 adult males (including 2 permanently attached workers) working full-time; 3 males (family members) working part-time; total labour-time 105 M.W.D. (i.e. 25+25+25+10+5+15); female labour is completely absent on this farm.

PROGRAMMING RESULTS

FARM No. 1

Income to	Crop Activities (acres)			Marginal Value Productivity (Rs.)		
Fixed Factors (Rs.)	*Cotton, K, A, a/s 100*	*Wheat-grain R mix 100*	*Wheat R mix 100*	*Labour October*	*Land Kharif*	*Land Rabi*
2995.53	8.60	8.38	.22	7.83	115.66	157.16

FARM No. 2

Income to	Crop Activities (acres)				Marginal Value Productivity (Rs.)			
Fixed Factors (Rs.)	*Cotton K A a/s 100*	*Paddy K a/s 100*	*Fodder K a/s 100*	*Wheat R mix 130*	*Labour June*	*Labour October*	*Land Rabi*	*Labour July*
2356.2	1.49	1.13	5.06	9.6	5.62	0.39	104.54	14.79

FARM No. 3

Income to	Crop Activities (acres)			Marginal Value Productivity (Rs.)		
Fixed Factors (Rs.)	*Crop K A a/s 100*	*Vegetables K mix 100*	*Wheat R mix 100*	*Labour October*	*Land Kharif*	*Land Rabi*
4156.88	5.72	5.78	12.50	8.5	109	158.95

FARM No. 4

Income to	*Crop Activities (Acres)*						*Marginal Value Productivity (Rs.)*				
Fixed Factors Rs.	*Cotton K A a/s 100*	*Maize K mix 100*	*Vegetables K,R mix 149*	*Fodder K,R mix 100*	*Wheat R mix 100*		*Labour Sept.*	*Labour Oct.*	*Labour Nov.*	*Land Kharif*	*Land Rabi*
4282.21	3.25	4.29	3.31	2.32	7.54		18.56	6.71	0.95	13.64	92.52

FARM No. 5 MODEL I

Income to	*Crop Activities (Acres)*				*Marginal Value Productivity (Rs.)*			
Fixed Factors (Rs.)	*Cotton K A a/s 150 irrigated*	*Bajra K FYM Irrig.*	*Wheat gram R mix 100 Irrig.*	*Wheat Unirrig. R*	*Labour Oct.*	*Land Kharif Irrig.*	*Land Rabi Irrig.*	*Land Unirri.*
4167.74	8.42	0.8	8.50	10.25	15.07	64.5	150.0	73.91

FARM No. 5 MODEL II

Income to	*Crop Activities (Acres)*			*Marginal Value Productivity (Rs.)*			
Fixed Factors (Rs.)	*Cotton K,A, a/s 150*	*Wheat gram mix 150*	*Fodder R*	*Labour July*	*Labour September*	*Labour October*	*Land Rabi*
4860.45	8.75	17.50	0.80	1.35	17.03	10.22	101.5

SELECT BIBLIOGRAPHY

INDIA : OFFICIAL PUBLICATIONS

Papers and Proceedings Connected with the Passing of the Deccan Agriculturists' Relief Act (1875-94), 2 Volumes

The Famine Commission Report, 1880

The Famine Commission Report, 1901

The Royal Commission on Agriculture, 1928, *The Report and the Evidence Volumes*

Review of Irrigation in India, 1911-12, 1916-17, 1917-18, 1936-37

Proceedings of the Meeting of Crops and Soil Wing of the Board of Agriculture, 1936 and 1937

Report on the Marketing of Wheat in India, 1937

The Famine Commission Report, 1945

Co-ordination of Agricultural Statistics, 1950

Agricultural Labour Enquiry, 1950-51, 1954, Delhi,

Symposium on the Determination of Costs and Benefits of River Valley Projects, 1953, 1954 New Delhi

All India Rural Credit Survey, 1954

Indian Crop Calendar, 1956, 3rd edn.

The Integrated Scheme of Rural Credit, 1958

The Second Five Year Plan, Planning Commission, New Delhi, 1956

Food Grains Enquiry Committee Report, 1957

Consolidation of Holdings, Methods and Problems, 1958, Planning Commission, New Delhi.

The Report of the Committee of the Panel of Land Reforms, 1959

Selected Plan Statistics, Planning Commission, New Delhi, 1959

Agricultural Legislation in India, Vols. 1-8, 1952-58, Delhi

Second Agricultural Labour Enquiry, 1956-57, 1960, Delhi

Proceedings of the Meeting of Crops and Soil Wing of the Board of Agriculture, 1960

The Third Five Year Plan, 1961, Planning Commission, New Delhi

Studies in Agricultural Economics, Vols. 1-3, 1954-60, Ministry of Food and Agriculture, Delhi

Punjab Board of Economic Enquiry, *Farm Accounts in the Punjab*, 1926-27

Agricultural Statistics of the Punjab, 1901-2 to 1935-6, ed. J.W. Thomas

Studies in the Economics of Farm Management, Punjab, Bombay, etc. 1954-55 and 1955-56

UNITED NATIONS

Report on the United Nations Conference on the Application of Science and Technology for the Benefit of the Less Developed Areas, Vol. III—*Agriculture*, New York, 1963

Food and Agricultural Organisation

BINNS, B.O., *The Consolidation of Fragmented Agricultural Holdings*, 1950

BROSSARD, D. B., *New Approach to Agricultural Credit: Supervised Credit*

Combined with Co-operatives and Agricultural Extension, 1964
BIGBY, M., *Co-operatives and Land Use*, 1957

Farm planning and Management : papers presented at the F.A.O. Development Centre on Farm Planning and Management, India, October 14, November 9, 1957, New Delhi, 1959
KGNATIEFF, V., *Efficient Use of Fertilisers*, 1949
KELLOGG, C. E., *Conflicting Doctrines about Soils*, 1957

Land Reform: Defects in Agrarian Structure as Obstacles to Economic Development, 1951
LOPEZ, P., *Principles of Land Consolidation Legislation*, 1962
MOELNAAR, A., *Report to the Government of India on the Appraisal of the Water Resources of the State of Ajmer*, 1955
——A., *Report on the Soil Erosion Problems of U.P.*, 1955
PARSONS, K. H., *The Owner Cultivator in a Progressive Agriculture*, 1958
Progress in Land Reform, 1st and 2nd Reports, 1954, 1956

RAO, T. S., *Guide to Methods and Procedures of Rural Credit Surveys*, 1962

SKOVGAARD, K., *The Consolidation of Fragmented Agricultural Holdings*, 1950
STEPHENS, C. G., *Soil Surveys for Land Development*, 1953
Soil Conservation : An International Study, 1948
Soil Erosion by Wind and Measures for its Control on Agricultural Lands, 1960

WHYTE, R. O., *Legume in Agriculture*, 1953
WILLIAMS, M. B., *Essential Considerations for Irrigation Development*, 1951

YANG, W. Y., *Methods of Determining the Costs of Using Alternative Types of Farm Machinery and Implements*, 1950
———, *Methods of Farm Management Investigation for Improving Farm Productivity*, 1958
——, *Farm Development in Japan*, 1962

Organisation for European Economic Cooperation (O.E.E.C.)
Manures and Fertilisers : Potential Progress in Europe, Paris, 1954
The Small Family Farm: A European Problem—Methods for Creating Economically Viable Units, Paris, 1959
The Mechanisation of Small Farms in European Countries, Paris, 1950
International Conference
Report of the Second Meeting of the Working Party on Rice Breeding, International Rice Commission, Rome, 1951
International Meeting on Infestation of Food Stuffs, *Summary Proceedings*, London, 1947

AUTHORS

Association of the Agricultural Economists, *Proceedings of the International Conferences*, 1955 and 1958
ARROW, K. J., "Alternative Approaches to the Theory of Choice in Risk Taking Situations," *Econometrica*, Vol. XIX, No. 4, 1951

——"Utilities, Attitudes, Choices : A Review Note" *Econometrica*, Vol. XXVI, No. 1, 1958

BANERJI, S., "Meteorology in Indian Agriculture," *Agricultural Situation in India*, August, 1961

BANSI, P.C., Future Feed, Seed and Wastage Rates," *Indian Journal of Agricultural Economics*, July-September, 1959

Baum, Heady, *et al.*, ed. *Capital and Credit Needs in a Changing Agriculture*, Iowa State University Press, 1961

BLACK, J. D., AND STEWART, H. L., *Report on the Research , Teaching and Public Administration of the Economics of Agriculture for India*, Delhi, 1954

BRAITHWAITE, R. B., *Scientific Explanation : A Study of the Function of Theory, Probability and Law in Science*, Cambridge University Press, 1953

BURNS, W., *Technological Possibilities of Agricultural Development in India*, Government Press, Lahore, 1944

DANIEL, E. F., *Farm Planning and Management*, Directorate of Economics and Statistics, Ministry of Food and Agriculture, India, 1960

DARLING, M., *The Punjab Peasant in Prosperity and Debt*, Oxford Press, Bombay, 4th Edition, 1947

———, *Report on Certain Aspects of the Co-operative Movement in India*, Planning Commission, New Delhi, 1957

DASGUPTA, S., *A History of Indian Philosophy*, Vol. V, Cambridge University Press, 1932

DAS GUPTA, A., "India's Cultural Values & Economic Development: A Comment," *Economic Development and Cultural Change*, October, 1964

DESAI, D. K., *Increasing Income and Production in Indian Farming*, Indian Society of Agricultural Economics, Bombay, 1963

DIGGS, B. J., "A Technical Ought," *Mind*, Vol. LXIX, 1960

DUBEY, V., "Marketed Agricultural Surplus and Economic Growth in Under-developed Countries," *Economic Journal*, December, 1963

ELLSBERG, D., "Risk, Ambiguity and the Savage Axioms," *Quarterly Journal of Economics*, LXXV, 4, 1961

EPSTEIN, S., *Economic Development and Social Change in South India*, University Press, Oxford, 1962

FELLNER, W., "Distortion of Subjective Probabilities as a Reaction to Uncertainty," *Quarterly Journal of Economics*, LXXV, 4, 1961

GOHEEN, J., SRINIVAS, M.N., KARUE, D. G., and SINGER, M., "India's Cultural Values & Economic Development : A Discusion," *Economic Development and Cultural Change*, October 1958

GOULD, G., and SHACKLE, G. L. S., "Odds, Possibility and Plausibility in Shackle's Theory of Decision : A Discussion," *Economic Journal*, Vol. LXVII, 1957

GUPTA, S. C., *An Economic Survey of Shamaspur Village*, Dist. Shamaspur, Uttar Pradesh, Asia Publishing House, Bombay, 1959

HABAKKUK, H. J., *American and British Technology in the 19th Century*, Cambridge University Press, 1962

HAAVELMO, T., "The Notion of Involuntary Economic Decisions," *Econometrica* Vol. XVIII, No. 1

HEADY, E. O., *Economics of Agricultural Production and Resource Use*, New York, Prentice-Hall, 1952

——, *Agricultural Adjustment Problems in a Growing Economy*, Iowa State College Press, 1958

——, "Diversification in Resource Allocation and Minimization of Income Variability," *Journal of Farm Economics*, 1952

———, *Agricultural Policy un ler Economic Development*, Iowa University Press 1962

———, AND CANDLER, W., *Linear Programming Methods*, Iowa State University Press, 1960

HILDRETH, C., "Problems of Uncertainty in Farm Planning, Discussion with D. A. Knight," *Journal of Farm Economics*, December, 1657

HOWARDS, A., and Howard G.L.C., *The Application of Science to Crop Production*, Oxford, 1929

HOWARD, G.L.C., *Sir Albert Howard in India*, Faber and Faber, London, 1953

INDIAN SOCIETY OF AGRICULTURAL RESEARCH, *The All India Soil Survey Scheme*, Bombay, 1953

———, *Indigenous Agricultural Implements of India : An All-India Survey*, Bombay, 1960

———, *Problems of Farm Costs in Indian Agriculture : A Study in Methodology*, Bombay, 1953

———, *Seminar on Rationale of Regional Variations in the Agrarian Structure of India*, Bombay, 1957

———, *Bhadkad : Social and Economic Survey of a village : A Comparative Study*, 1915 and 1955., Bombay, 1957

JENNY, H., *Factors in Soil Formation*, McGraw Hill, New York, 1941

———, and RAY CHOUDHURI, S. P., *Effect of Climate and Cultivation on Nitrogen and Organic Matter Reserves in Indian Soils*, Indian Council of Agricultural Research, New Delhi, 1960

JOHNSON, L. G., "Methodology for Studying Decision Making," *Journal of Farm Economics.*, December 1957

JOHNSTON, J., "Decision Theory," *Progress*, Spring 1963

KNIGHT, F. H., *Risk Uncertainty and Profit*, London University: reprints of "Scarce Tracts in Economics and Political Science", 1957

KRISHNA, R., "Farm Supply Response in India-Pakistan, A Case Study of the Punjab Region," *Economic Journal*, September 1963

LOVEDAY, A., *The History and Economics of Indian Famines*, Cambridge, 1914

LUCE, R. D., AND RAIFFA, H., *Games and Decisions*, John Wiley, New York, 1958

MAHALANOBIS, P. C., *Talks on Planning*, Indian Statistical Institute, 1961

MANN, H., *Rainfall and Famine — A Study of Rainfall in the Bombay Deccan*, 1865-1938, Indian Society of Agricultural Economics, Bombay, 1955.

MARKOWITZ, H. M., *Portfolio Selection: Efficient Diversification of Investments*, John Wiley and Sons., New York, 1955.

McFarquhar, A. M., "The Practical Use of Linear Programming in Farm Planning, *The Farm Economist*, Vol. IX, No. 10, 1961

———, "Rational Decision Making and Risk in Farm Planning: An Application of Quadratic Programming in British Arable Farming," *Journal of Agricultural Economics*, Vol. XIV, No. 4, 1962

Mellor, J. W., and Stevens, R. D., "The Average and Marginal Product of Farm Labour in Underdeveloped Economics," *Journal of Farm Economics*, August 1956

———, "Increasing Agricultural Production in Early Stages of Economic Development: Relationships, Problems and Prospects," *Indian Journal of Agricultrual Economics*, April-June 1962

———, "The Use and Productivity of Farm Family Labour in the Early Stages of Agricultural Development," *Journal of Farm Economics*, August 1963

Moglewer, S., "A Game Theory Model for Agricultural Crop Selection," *Econometrica*, Vol. XXX, No. 2, 1962

Mukherjee, P. K., *Economic Surveys in Underdeveloped Countries — A Study in Methodology*, Asia Publishing House, Bombay, 1959

———, *et al., A Pilot Survey of the Fourteen Villages in the U.P. and the Punjab*, Asia Publishing House, Bombay, 1959

Narain, D., *Distribution of the Marketed Surplus of Agricultural Produce*, Asia Publishing House, Bombay, 1961

Nerlove, M., *The Dynamics of Supply: Estimation of Farmers' Response to Price*, John Hopkins Press, 1958

Nowell-Smith, P. H., *Ethics*, Penguin, 1954

Ogg, W. G., "*The Role of Fertilisers in the National Economy*," *Rothamsted Memoirs on Agricultural Chemistry and Physiology, 1945-48*

Ojala, E. M., *Agriculture and Economic Progress*, Oxford University Press, 1952

Ohkawa, K., and Rosovsky, H., "The Role of Agriculture in Modern Japanese Economic Development," *Economic Development and Cultural Change*, October, 1960

Popper, K. R., *Conjectures and Refutations: The Growth of Scientific Knowledge*, Routledge and Kegan Paul, London, 1963

———, *The Logic of Scientific Discovery*, Hutchinson, London, 1959

Panikar, P.G.K., "Rural Savings in India," *Economic Development and Cultural Change*, October 1961

Raj, K. N., *Some Economic Aspects of the Economics of Bhakra Nangal Project: A Preliminary Analysis in terms of Selected Investment Criteria*, Asia Publishing House, Bombay, 1960

Rao, V. K. R. V., Sen, S. K., *et al., Papers on National Income*, Asia Publishing House, Bombay, 1960

Raiffa, H., "Risk Ambiguity and the Savage Axioms," *Quarterly Journal of Economics*, Vol. LXXV, No. 4, 1961

RENBORG, ULF, *Studies on the Planning Environment of the Agricultural Farm*, Uppsala, 1962

REDDAWAY, W. B., *The Development of the Indian Economy*, George Allen and Unwin, London, 1962

RUSSELL, J., *Report on the Working of the Imperial Council of Agricultural Research in Applying Science to Crop Production in India*, Delhi, 1937

SAVAGE, L. J., *The Foundations of Statistics*, John Wiley and Sons, New York, 1954

SCHICKELE, R., "Farmers' Adaptation to Income Uncertainty," *Journal of Farm Economics*, August 1950.

SHACKLE, G. L. S., *Expectations in Economics*, Cambridge University Press, 1949

SHAH, C. H., *Problems of Small Farmers, Report on an Enquiry into the Problems of Low-Income Farmers in Kodinar Taluka*, Indian Society of Agricultural Economics, Bombay, 1958

SHARPE, W. F., "A Simplified Model for Portfolio Analysis," *Management Science*, Vol. IX, No. 2, 1963

SHEPHERD, G. S., *Agricultural Price Analysis*, Iowa University Press, 1941

SHUBIK, M., "Information, Risk, Ignorance and Indeterminacy," *Quarterly Journal of Economics*, Vol. LXVIII, No. 4, 1954

SCHULTZ, T., *The Economic Organization of Agriculture*, McGraw-Hill, New York, 1953

———, *Agriculture in an Unstable Economy*, New York, 1945

SIEGEL, S., "Level of Aspiration and Decision Making," *Psychological Review*, Vol. LXIV, No. 4, 1957

SIMON, H. A., "Theory of Decision Making in Economics and Behavioural Science," *American Economic Review*, Vol. XLIX, No. 3, 1959

———, "A Behavioural Model of Rational Choice," *Quarterly Journal of Economics*, Vol. LXIX, No. 1, 1955

SINGERS, C., et al., *A History of Technology*, Oxford, 1954-1958, Vol. IV and V

STEWART, A. B., *Report on Soil Fertility Investigations in India with Special Reference to Manuring*, Army Press, Delhi, 1947

TURVEY, R. GRAAF, J. DE. V., BAUMOL, W. J., and SHACKLE, G. L. S., "Three Notes on Expectation in Economics," *Economica*, Vol XVI, 1949

THORNER, DANIEL AND ALICE, *Land and Labour in India*, Asia Publishing House, Bombay, 1962

VOELCKER, J. A., *The Report on the Improvement of Indian Agriculture to the Government of India*, London, 1893

WALTER-LEATHER, J., *Final Report to the Government of India*, Dehra Dun, India, 1897.

WARRINER, D., *The Economics of Peasant Farming*, London, 1939

———, *Land Reform and Development in the Middle East*, London, 1957

WHITE, L. T., *Medieval Technology and Social Change*, Clarendon Press, Oxford, 1962

WILLIAMS, J. D., *The Complete Strategist*, McGraw Hill, New York, 1954

INDEX